FINAL
SOLUTIONS

FINAL
SOLUTIONS

BIOLOGY, PREJUDICE, AND GENOCIDE

RICHARD M. LERNER

WITH FOREWORDS BY
R. C. LEWONTIN
AND
BENNO MÜLLER-HILL

THE PENNSYLVANIA STATE UNIVERSITY PRESS
UNIVERSITY PARK, PENNSYLVANIA

Library of Congress Cataloging-in-Publication Data

Lerner, Richard M.
    Final solutions : biology, prejudice, and genocide / Richard M.
Lerner ; foreword by R. C. Lewontin, foreword by Benno Müller-
Hill.
        p.   cm.
    Includes bibliographical references and index.
    ISBN 0-271-00793-1
    1. Sociobiology. 2. Biology—Social aspects. 3. Social policy.
4. Nature and nurture. 5. National socialism. 6. Lorenz, Konrad,
1903–1989. I. Title.
GN365.9.L47   1992
304.5—dc20   91-19711
                                                                          CIP

It is the policy of The Pennsylvania State University Press to use acid-free paper for
the first printing of all clothbound books. Publications on uncoated stock satisfy the
minimum requirements of American National Standard for Information Sciences—
Permanence of Paper for Printed Library Materials, ANSI Z39.48–1984.

CONTENTS

FOREWORD

by R. C. Lewontin

As the reader of this book begins to turn its pages, Europe is in the process of turning back the pages of history. As a consequence of the failure of socialist governments to solve the social problems that have faced their nations, a new life has been given to the most retrograde and destructive aspect of European consciousness, national chauvinism. The narrow-minded and violent nationalisms that characterized Europe before the Second World War seemed, on the surface, to be waning as a result of state policy, not only in socialist countries but in capitalist countries as well. Yet, underneath, they have remained and have now broken out with renewed fury. The people of the Baltic, the Balkans, the Caucasus through Turkey into the Middle East are killing each other in the name of national identity. In capitalist countries too, despite the economic internationalism of state policy reflected in the European Economic Community, Basques are blowing up Spaniards, a large party in France is demanding the expulsion of all "foreigners" (Algerians), working-class English whites are bashing working-class English Pakistanis and blacks. Lithuania for the Lithuanians, Kurdistan for the Kurds, Georgia for the Georgians, France for the French, any-land for the any-landers—the task of filling in the blanks is left to the reader.

Sometimes the demand for national self-determination is a response to real exploitation by external forces. "Nicaragua for the Nicaraguans" was a reaction to the reality of North American economic and political hegemony that kept Central America in a position of dependent poverty. Sometimes nationalisms are the deliberately created false consciousness that "foreign" minorities within the state are responsible for the unsatisfactory state of people's lives, as in the long-standing anti-Semitism of Europe. But whatever the generating forces that keep nationalism alive, however true or false their identification of exploitative groups, they must, in the end, assert the unchanging and unchangeable nature of social identity. "Once a Kurd always a Kurd" and "Once a Jew always a Jew" are slogans not only of Turks and Nazis but of Kurds and Jews as well. Exploiters and exploited alike share in the consciousness of a cultural and biological heritage that marks out indelible group boundaries that transcend human historical development. By the same mysterious process of biological and deep cultural influence, we are ultimately what our ancestors made us, try as we will to become something different. In his book on the Marranos, Cecil Roth, the great authority on the history of the Jews, relates the case of a Jewish family who arrived in England in the early eighteenth century and who, by a series of marriages, integrated into the English landed aristocracy, until one of its members became Chancellor of the Exchequer in a Tory government. But, you see, Roth tells us, race will tell, because, after all, it was Chancellor of the *Exchequer*! Indeed, Roth in his determinism is very different from the officers of the Inquisition, who, in common with the propagators of that other evangelical faith, Islam, thought that people could change if only they got their heads straight.

Final Solutions is a book about the extreme horrors that arise when people take seriously the proposition that there are racial and national characters from which we cannot escape. Unlike other books on Nazi race theory, *Final Solutions* extends ideas of determinism beyond the purely biological into the cultural as well. Cultural determinism—the doctrine that our cultural heritage passed down by a process of unconscious acculturation is inescapable—differs only in a trivial mechanical detail from biological determinism, the doctrine that we cannot escape our genes. Both biological and cultural determinism deny essential freedom to human consciousness.

A great deal of attention has been given to debunking biological determinism by showing that it is simply bad biology. It supposes that genes control consciousness, making us more or less entrepreneurial, aggressive, clever, or religious, whereas there is no evidence at all that genes can act in that way. Moreover, biological determinism assumes that genes do not simply influence but *determine* our beings. It makes the error of equating heritable with unchangeable, a biological mistake of the first magnitude. Many works have been written (I have written some myself) explaining why this is pseudo-scientific nonsense. The possibility of revealing the errors of biological determinism has rested on the very great knowledge we now have of the relations between genes and development. So we can say with certainty that one or another claim of biological determinism is false.

Much less attention has been given to an analysis of cultural determinism, which is sometimes proposed as the only alternative to biology. After all, if our genes do not make us what we are, then our environment must. Partly, then, the lack of analysis of environmental determinism has been a consequence of the belief that it is the only alternative to biologism. But, in addition, we have very little solid information about how "culture" influences the individual in the same sense that we have solid information about how genes influence development. Theories of culture are speculative and ideological, and nothing is obviously true, so cultural determinism is a more slippery concept. At base, however, it depends on an alienated view of organism and environment, about which we do know something.

Cultural determinism assumes that there is a fixed external world, culture, and that individuals react to it and are molded by it. That is the same view as the adaptationist stance in evolutionary biology. But in the biological world, organisms do not simply react to an autonomous external nature, but actively form nature. Just as there is no organism without an environment, so there is no environment without an organism. The relation between organism and environment is a dialectical one of co-determination in which both organism and environment are subject and object. So, too, human culture is not autonomous but is in constant process of historical development by the action of individuals who are, in turn, formed by culture. And, for an individual life, there is superimposed the freedom of individual decision and action that are constrained but not determined by external circumstances.

This dialectical relation is what Richard Lerner calls "developmental contextualism." It is the alternative to biological and cultural determinism. It is the statement of the developmental contextual view that is the important central point of *Final Solutions*, and it is the full elaboration of that point of view that is a pressing program for social theory. Nowhere has this world view been put more succinctly than in Marx's third Thesis on Feuerbach:

> The materialist doctrine that men are products of circumstance and upbringing, and that, therefore, changed men are products of other circumstances and changed upbringing, forgets that it is men that change circumstances and that the educator himself needs educating.

Cambridge, Massachusetts
July 1991

FOREWORD

by Benno Müller-Hill

The Nazis would not have achieved power if their general ideology had not been widely accepted in Germany. They could not have realized their racial program—including the murder of Jews, Gypsies, and the insane—without the help of anthropologists, geneticists, psychiatrists, and other scientific experts. It took a terrible war to make the Nazis disappear. But did their ideology disappear with them? What was their ideology?

Richard Lerner identifies biological determinism as the central dogma of the Nazi ideology, or religion. Biological determinists claim that, to a large extent, genes determine human behavior. The variant genes that win in the struggle for survival are the better ones, they say, and thus even ethics is determined by evolution. It is the strategy of the winner. But, if this is true, then analyzing the behavior of ants, bees, or ducks can bring us to a true understanding of human behavior. A witty molecular biologist once coined the phrase "What is true for *E. coli* is true for the elephant." The devoted human sociobiologist could therefore say: "What is true for the duck is true for humankind."

So I particularly like the chapter on Konrad Lorenz, the Austrian Nobel Prize winner. Between 1940 and 1943 Lorenz designed a form of the central dogma of the Nazi religion which was able to survive the defeat of the Nazis. How did he do that? He never called a Jew a Jew; he just called the Jews and all the others "the stigmatized" *(die Gezeichneten)*. He never wrote that "the stigmatized" should be killed; he just wrote that they should be eliminated by those who had now replaced the selecting forces of evolution. If his science was good enough for a Nobel Prize, what can be wrong with his ideology? Lerner cites a blunt letter to the editors of a journal in 1976, in which Lorenz points out that he was never inclined against the "ethnically inferior," just against the "ethically inferior." What a fine distinction!

Who are the superiors, who are the inferiors? Who are the winners, who are the losers? One hundred years ago a great many biologists saw Caucasians as the winners and Blacks as the losers. The Nazis added the Jews as the main losers among the Caucasians. Slaves in Egypt, and homeless for two thousand years in Europe and elsewhere, the Jews were destined, in the view of the Nazis, to become the total losers.

The view that all human behavior is essentially determined by genes and that the particular variants we have are determined by natural selection is a religious statement not a scientific statement. Thus, science here does what it should never do: it makes value judgments. That this kind of science is to some extent pseudoscience is a particular irony, but that does not provide much relief. As a scientist I dislike intensely the notion that some Christians want to teach creationism instead of evolution. But as a citizen I dislike equally that scientists advocate the religion of science in their classrooms when they teach human sociobiology. There should be a clear division between religion and science in schools, colleges, and universities. The old battle cry of Voltaire, "Écrasez l'infâme," should be voiced today against the religion of science and particularly the dogma of biological determinism.

Lerner spends a good deal of time demonstrating that today's sociobiology uses an obsolete notion of the heritability of intellectual capabilities. This is true as long as only the phenotype is analyzed. But that is going to change. In the next years and in the coming century, it will be possible to juxtapose human phenotype (behavior) with human genotype (DNA sequence). This will change human genetics radically. With the Human Genome Project advancing, this type of analysis will become better and better. Never mind that some, or perhaps even many, of the attempts to do this today will then be shown to be wrong. The RFLP analysis of today will be obsolete when one can look at real genes.

This is just a question of time. All farsighted scientists should have an interest here. If they yield to the religious zealots of biological determinism, they will help to create a world that will rival that of the Nazis. Scientists who allow this to happen will lose all credibility. Thus, those of us who are against the abominable religion of biological determinism should know how to defend our view.

Richard Lerner's case for the effects of the environment, which he calls developmental contextualism, is such a defense. But one should not create false hopes. Those who have a child with Down's syndrome are badly served with wrong hopes; they need financial help from society. Free choice and civil rights will have to be defended. Special legislation will be necessary to protect the losers. But that is a different story, and one for the next century.

Cologne
May 1991

PREFACE

As a Jewish boy growing up in Brooklyn in the late 1940s and early 1950s I could not escape Hitler. He, Nazis, the Gestapo, Auschwitz were everywhere: in the tears in my aunts' eyes as they looked at pictures of their siblings who did not survive; in the tattooed numbers on the wrists of my uncles who did; and in the countless stories told by my grandmother, who wanted to be sure we would always know who we were.

"You once had a Cousin Maitka," Grandmother told me. She lay next to me on a bed in the spare bedroom of her Brownsville apartment. She stroked my head to—I thought—push me closer and closer to the nap I was trying not to take.

"What happened to her, Grandma?"

"She died." The response came softly but quickly. Her voice cracked the haze of the early afternoon sun filtering through the venetian blinds.

"How'd she die?" I gasped it out. I was not prepared for the answer I heard.

"Hitler's soldiers killed her. She was your Aunt Helen's baby. She was six months old. Your aunt tied her to a tree—with a potato in her mouth. When the soldiers found her they killed her."

Sleep was now far off. "Why would a mommy tie up her baby so she would be killed? Why would a mommy do that?" I was confused, afraid. This was not part of the world I knew.

"Haschka had no choice," she explained, now using Helen's Yiddish name. "Haschka, her brothers Herschel and Israel, and the other people from their village were hiding in the woods. They were all hiding together. They were trying to be very quiet. If the Nazis heard them, they'd all be caught and shot. They thought if they could get through the woods they'd be safe. But the Nazis were coming closer. Haschka's daughter was the only baby with them. All the grown-ups could be quiet, but if the baby cried that would let the Nazis know where everyone was. So everyone said Haschka had to make sure the baby wouldn't cry. The only way she could do that was to put a potato in her mouth and tie her to a tree so she couldn't knock the potato out or crawl away. Then everybody ran through the woods to escape."

"What happened to the baby then?" I was terrified.

"No one knows. Haschka hopes someone found her and saved her—that they kept her as their own baby."

"But what if they didn't?"

"Then she's dead. She is dead." She said this in a matter-of-fact way in a resigned, hollow tone. This was the way of her world. She wanted me to know this world, to feel it, and in her own way to so imbue me with it that I would learn never to live in it.

"But, Grandma, why'd Haschka do that? Couldn't she just keep her hand over the baby's mouth or something?"

"No. There was no other way. Either Haschka had to give up her baby or everyone would have been shot. The Nazis would have killed everyone."

"Why, Grandma, why?" I was just about crying now. Big tears welled up in my eyes.

"Because," she said softly and now consolingly, "because they were Jews. And Nazis hate Jews."

"Why?" I was almost pleading now. "Why do they hate Jews?"

"Just because. Just because. If you're Jewish, Nazis will try to kill you. They just hate you because you're Jewish."

My last question, more shuddered than spoken, came out: "Will it happen to me? Will they try to kill me too?" I began to cry.

"Hitler's dead now" was all she said. "Hitler's dead now."

I pressed as close to her as I could. She hugged me close, tightly— probably, as I now look back on it, as she imagined all the mothers who had lost all those children had hugged them as they took their final steps—to stand in the execution pits of Babi Yar before the firing squad or to huddle behind the iron doors of the gas chambers of Auschwitz and feel the steam of the Zyklon-B fill their throats.

I closed my eyes tightly, trying to escape from the marks my grandmother's words had made on me. Yet I could not help seeing the big eyes of a little girl I'd never known open wide to greet a man walking toward her with his rifle pointed right at her head. . . .

In the time that has passed since that afternoon in my grandmother's apartment I have learned—and increasingly so as the years go by—how deeply I was affected by these early lessons about Nazi genocide. I now understand that much of my life has been shaped by my attempts to go beyond the answer of "Just because." In fact, perhaps my wise and lovingly manipulative grandmother gave me an unsatisfying answer in hopes of motivating just such a search.

My attempts to find a better answer have been a central part of my professional and scholarly career. My search led me to study psychology, through which I began to understand why people think and behave in particular ways. As I learned more, I was drawn to the area of developmental psychology, where a central issue—the nature-nurture controversy—was associated with a literature that went beyond my grandmother's answer "Just because." I learned that some scholars believed that people developed their specific behaviors, thoughts, or feelings primarily because of their biology or, more specifically, their heredity, their nature. Others accounted for development by reference to primarily environmental or experiential (nurture) factors. Still others believed that nature and nurture might interrelate to account for a person's actions, thoughts, and emotions.

As I read the literature on human development and began contributing to the debate on nature versus nurture, I soon discovered a way to go beyond my grandmother's answer. Indeed, I have detected a common theme that emerges from virtually all my empirical research and theory: Neither biology nor environment alone is sufficient to account for human behavior and development; instead, behavior and development derive from a complex, dynamic interrelationship between, or fusion of, nature and nurture.

The belief in such a dynamic relationship has never been a majority view in either the biological or the behavioral/social sciences. In different disciplines and/or at different points in time, one or the other has been predominant. Environmentalists believe the causes of behavior lie in "nurture"—in the context surrounding an organism. Thus, they maintain views that stress nurture over nature—that is, environment over biology. Although environmentalists see biology as necessary for placing an organism in the context, they believe the major causes of behavior and development lie in the environment. Important exceptions notwithstanding (for example, involving "nature"-oriented psychologists, such as William McDougall), American academic psychology through much of the twentieth century focused most often on such an environmentalist view. Similarly, at this writing, major theoretical themes in sociology and anthropology continue to be based on sociogenic or culturally deterministic interpretations of individual behaviors and of social structure and change.

An alternative to environmental or cultural determination has been the doctrine of biological determination, of nature over nurture. In the latest form of this doctrine, for instance, the primary cause of behavior lies in the evolutionarily shaped, hereditary processes of individuals. The basis of a person's behavior and development is said to lie in the genes. And because society is nothing more than a collection of individuals, evolutionary-based hereditary mechanisms create and drive society too. The doctrine of biological determinism can be traced at least to the writings of Plato. One or another minor variant of the key idea within this doctrine—that human behavior and development can be reduced to genetic processes—has shaped major ideas in biology, medicine, and social and behavioral science. Freud's psychoanalytic theory of development, the nineteenth- and early twentieth-century embryological work of Ernst Haeckel, the ideas of the European and American Social Darwinists of the nineteenth and twentieth centuries, the American and European eugenics movement during the same period, the German racial hygiene movement of the first half of the twentieth century, and the contemporary "synthetic" science of sociobiology are all positions shaped at their core by the doctrine of biological determinism.

Scientists may put the doctrines of biological and environmental determinism to different uses, employing ideas from one or both doctrines in whole or in part to organize or test ideas or hypotheses. What is of concern to me here, however, is when one of the doctrines is used as a pervasive *and implicit* organizing orientation in a scientist's ideas and research. When doctrine is employed in such a way it becomes more of a presupposition about the world—that is, an a priori accepted and untested set of ideas about what is the case, rather than a conceptual framework for beginning to learn things about the world, for investigating and determining what is or may be the case. The use of such doctrine in this way—as equivalent to an ideology—is the concern of this book.

Although in this book I speak about the doctrine of environmental determinism and the doctrine of biological determinism, I focus on biological determinism. Scientists associated with both types of determinism have had their ideas translated into social policies, and sometimes those policies promoted inhumane treatment of human beings. However, it is the social policies derived from biological determinism that have most often directly determined whether people survived or not. Some of the greatest inhumanities humans have perpetrated on other human beings have been predicated in great part on the doctrine of biological determinism.

Biological determinists—scientists and politicians alike—believe that human behavior has been shaped by evolution to take its present form and that the immediate determinants of that form are genes. In addition, they believe that *differences* among people are based on genetic differences. Given this belief, they also hold that changing the environment cannot alter this connection between genes and behavior. Such a belief may not be harmful in and of itself. The problem comes when that belief becomes the scientific basis for specific social policies. If we do not appreciate or value the differences in behavior, thoughts, or values, we might think it necessary to do something to circumscribe or delimit people who are different from us. The reasoning would be that, because changing the environment cannot change behavior, we must take steps to protect ourselves from people who are "different," to ensure that neither those people nor their genes can diminish our own, valued behaviors.

Such reasoning—the reasoning of the ideology of biological determinism—can open a Pandora's box of delimiting, degrading, and finally fully dehumanizing social policies. Laws curtailing freedom of movement, or rights of property or citizenship, or freedom of marriage might be enacted; the Nuremberg laws of Nazi Germany and the apartheid policies in South Africa are cases in point. But even stronger measures might be taken if the threat to society's safety posed by the supposedly undesirable behaviors of those who possess the "wrong" genes was perceived to be great enough. Some very unique solutions, some quite special treatment—*Sonderbehandlung*, to use the Nazi term for such severe actions—might be deemed necessary: collection, or "concentration," of the undesirables, and perhaps then some ultimate, or final, solution (*Endlösung*) to ensure permanent protection against the threat of the undesired genes.

The Nazi program of genocide is clearly the most evil implementation of the ideology of biological determinism known in recorded history, and a major focus of this book is the connection between this ideology and the social policies of the Nazis. However, contemporary scholarship in biology and in the behavioral and social sciences is currently debating a new version of biological determinist ideology: sociobiology. Therefore, another focus of this book is the core conceptual relationship between the contemporary sociobiological version of biological determinism and the version of this ideology that underpinned Nazi social policy, a version involving ideas of Social Darwinism and of racial hygiene. I argue that both in general terms and in specific characterizations of particular groups of people—specifically Blacks and women—the Nazi racial hygienists and contemporary sociobiologists use identical concepts and strikingly comparable terminology—and draw consistently similar implications for social policy.

It is important to state at the outset that no contemporary sociobiologist recommends anything even remotely comparable to the "special treatment" or "final solution" of the Nazis. More important, there is no inevitable connection between a biological determinist scientific theory and a racist political ideology that uses biological ideas to legitimate the planks of its political platform. However, biological determinist theory purports to be based firmly on scientific evidence that there are genetically fixed and immutable differences between groups of people, and racist politicians are psychologically prepared to adopt such ideas in defense of their own agenda. Therefore, when a coupling of biological determinism and political racism occurs—as it inevitably does in fascist political movements—the effect is the apparent scientific legitimation of policies that constrain and delimit the opportunities and life chances for the people thought to have inherited the inferior or problematic genes.

In rejecting biological determinist ideology, one need not deny the contribution biological factors make to human behavior and development, or fall into an equally implausible (but perhaps less often socially pernicious) environmental determinism. An alternative view, the one I noted earlier, is that biology and environment, "nature and nurture," may be understood as fused across life, as being in a dynamic interrelationship. A third focus of this book is the scientific usefulness of this view of biology, context, and behavior. Known in my field of specialization as "developmental contextualism," this position is derived from the "levels of integration" concept of the comparative psychologist T. C. Schneirla. My presentation of developmental contextualism involves both an explication of its scientific characteristics and an indication of the implications of this view for social policy.

By examining biological determinism and its contemporary sociobiological version in relation to the social policies of the Nazis, and by looking at the scientific usefulness of "developmental contextualism," I hope to go beyond my grandmother's "Just because" and show how awareness of biological determinist ideology and its implications can inform us about why certain events took place in the past and protect us against a future in which actions based on biological determinist ideology might be repeated.

I owe great debts to many people for aiding me in writing this book. Perhaps my most general intellectual debt is due to someone I never met, the late comparative psychologist T. C. Schneirla. His theory, research, and social awareness about the dangers of the doctrine of biological determinism have been central in shaping my own work and in influencing my views about the threats imposed by this doctrine. Schneirla's influence permeates the pages of this book. So too, however, does the work of other scholars. Ethel Tobach has been a leading spokesperson for the view of nature-nurture fusion I present in this book and a voice of warning about the dangers of political co-optation of biological determinist ideology. Her influence also is pervasive throughout the book. I have relied heavily on the contributions of other scholars as well: R. C. Lewontin, Steven Rose, and Leon Kamin; Robert Proctor; Robert Jay Lifton; Stephen Jay Gould; Stephen L. Chorover; Theodora Kalikow; Lucy Dawidowicz; Benno Müller-Hill; and Gilbert Gottlieb. Some of these scholars I do not know personally, but the work of all of them has guided and encouraged me in the directions I followed.

My colleagues and students at Penn State, and my colleagues throughout the United States and abroad, have given me more-direct guidance and support, through their discussions and critiques of all or part of this manuscript. I wish to give my sincere thanks to Marc Bornstein, Roger A. Dixon, Lawrence Erlbaum, Alexander von Eye, David L. Featherman, Nancy Galambos, Miles Huston, Michael Lamb, Jacqueline V. Lerner, Michael Lewis, Patty Mulkeen, John R. Nesselroade, Jacqueline Schwab, Robert A. Scott, Alexander W. Siegel, Rainer Silbereisen, and Fay Wohlwill. A very special note of thanks goes to Teri Charmbury, who served as both typist and copy editor for the book and was as well an astute and helpful critic. I am also grateful to R. C. Lewontin and to Benno Müller-Hill for being generous enough to write forewords to this book.

I wish to express my appreciation also to my agent, Gloria Stern, whose support and faith in the book were welcomed and vital ingredients for its completion. In addition, I am grateful to my editor at Penn State Press, Peter J. Potter, for his enthusiasm about the book and for his superb and sage editorial advice. I also thank Peggy Hoover, of Penn State Press as well, for her insightful and always helpful editorial advice and guidance.

As always, I express my greatest debt to my family, to my wife and friend Jackie, and to our children, Justin Samuel, Blair Elizabeth, and Jarrett Maxwell. Their love, support, tolerance, and (when needed) distraction helped keep me going while reading and writing about bleak and frightening times, places, and concepts.

Finally, I dedicate this book to my family—and not just to my wife and children but to all my relatives—those whom I have known and loved throughout my life and those whom, because of their presence in another place at another time, I never had the opportunity to know. Your lives will not be forgotten.

RICHARD M. LERNER

1

NATURE, NURTURE, AND SOCIAL POLICY

A devil, a born devil, on whose nature
Nurture can never stick.
    —*Shakespeare,* The Tempest

The facts of science affect people's lives. These facts may enhance the ease of our existence—for example, through the continuing invention of a plethora of electronic devices. These facts also may threaten our existence—for example, through the too facile development and deployment of nuclear arms. It is equally true that scientific beliefs or philosophies influence people's lives. For instance, the belief that mental abilities can be represented by a single score from an IQ test altered U.S. immigration laws beginning in the early decades of the twentieth century and kept thousands of Eastern Europeans from entering the United States. Similarly, the theory that mental illness is caused by biochemical dysfunctions occurring among nerve cells led many psychiatrists to surgically remove portions of the brains of mental patients. More recently, this same theory has resulted in reemergence of electroshock as a treatment to "restir" the chemicals of the brain.

PHILOSOPHICAL MODELS OF HUMAN DEVELOPMENT

For at least the past twenty years a major interest of social and behavioral scientists has been how philosophical ideas about the basis of human development influence scientific theory and method.[1] One philosophical model views people as passive and machine-like and seeks to reduce human life to small (molecular), constituent characteristics or events; in some theories associated with this *mechanistic* model, these characteristics are construed to be connections between environmental stimuli and a person's responses. In other mechanistic theories these elements are conceived of as the hereditary units, or genes, given each person at conception.[2]

Another philosophical model with which I have been concerned sees humans as active, organized wholes. Theories associated with this *organismic* conception posit that, although a living organism is made up of different parts (a heart, a brain, a liver, etc.), any attempt to understand life through study of a part in isolation will be inadequate because, it is believed, when the parts combine to constitute a whole organism they produce, in their *combination*, characteristics that do not exist in any single part in isolation. For instance, the specific personality of individuals, or their abilities to love, hate, deny themselves for the sake of their children, or sacrifice themselves for the good of their values or for patriotism, cannot be understood by reducing the study of a person to a brain or a heart or a gene alone.

Organismic theorists also believe that these "emergent" characteristics of people, which occur as the parts are organized into a whole, change across life. They believe that there are universal sequences, or stages, of change in the structure or organization of a human being's life[3] and that therefore the qualities that define the whole—the person—will change across the span of the individual's development.

A third philosophical model conceives of humans as active and selective, engaged in dynamic relationships with a complex and changing world. Associated with studies of animal behavior and human development, this conception emphasizes that biology develops in relation to the complex and changing contexts of social life. Biological processes, within the person, are seen to develop in relation to social contexts outside of that person, contexts such as the family, the community, and the culture or environment. The emphasis on development and context is why this view is labeled *developmental contextualism* in modern psychology. Developmental contextualism has helped psychologists understand why a common biological change—for example, the beginning of a female's menstrual cycle in early adolescence—may result in fears and psychological problems for some girls *or* a sense of accomplishment, maturity, and pride in others. Reasons for such differences lie in the distinct family, peer group, and educational experiences encountered by the two types of girls.[4] Similarly, the mean age of menarche has been dropping by about four months a decade since about the mid-nineteenth century; girls have simply been starting their menstrual cycles at earlier and earlier ages.[5] Developmental contextualism has helped scientists understand that this association between biological change and *history*— the most general instance of the human context—is linked to improvements across time in nutrition and systems of health-care delivery.[6]

Thus, by stressing that different integrative relationships, or fusions, exist between biology, psychology, and context, ideas associated with developmental contextualism help scientists understand that biology is part of a process that is liberating human life, not constraining it.[7] The fusion between biological and contextual change means that human life is potentially more variable, more *plastic*, than other views of human development—that is, mechanistic or organismic views—hold. In other words, the developmental contextual perspective demonstrates how human existence can be improved, and in fact "recreated," across the entire span of human life as a consequence of biology-person-context interrelationships.

Over the course of the last decade, it has become clear that these different philosophical models—mechanism, organicism, and developmental contextualism—are associated with distinct theories of the course of human life and with diverse agendas for research across the life span. Most important, it has been demonstrated that the different philosophical models about human development lead to significant distinctions in theories about the basis of human development.

Mechanistic conceptions see human life as directed by forces one cannot control, forces that are placed either *onto* or *into* the person. The former, extrinsic, type of forces, those placed onto the human being, are the environmental stimuli just noted. Such forces involve what is commonly called nurture—socialization, associative learning, conditioning, and the like. The belief in such forces led B. F. Skinner to claim, in his 1971 book *Beyond Freedom and Dignity*, that "a person does not act upon the world, the world acts upon him."[8] The latter forces, those intrinsic forces placed into the person in another expression of the mechanistic conception, stress nature: genes and the mechanisms of heredity. Although genes exist within a person, a person's actions can no more influence which genes he or she inherits at conception than the person can, in Skinner's view of the world, shape his or her environment. Human development in mechanistic conceptions involves behavior that is either compelled by our nurture or predestined by our nature.

Many organismic conceptions of the basis of development are not radically different from the mechanistic ideas, which stress nature over nurture.

In the classical organismic conception, nature (e.g., genes or maturation) predetermines the sequence of changes through which a person progresses. For example, in Sigmund Freud's theory of psychosexual development all normal people go through the oral, anal, phallic, latency, and genital stages of development because these stages are universal and biologically predetermined. Nurture—environment—is said to interact with these biologically determined changes, but only as a secondary influence; environment may only speed up, slow down, or in extreme cases arrest these intrinsic progressions. However, nurture, seen as something separate from nature, cannot alter the sequence or focus of any of the stages.[9]

In my view, both mechanism and organicism are no longer prominent influences in contemporary research in human development. Mechanistic ideas have resulted in theories that are regarded as too reductionistic, that are seen as too concerned with interpreting qualitative distinctions between people and their contexts as being constructed of an all-determining set of common elements, be they genes or simple stimulus-response connections. In turn, organismic ideas have resulted in theories that are seen as too general to depict usefully the considerable individual distinctiveness of the lives of different human beings. That individuality—which tends to *increase* across the life span, and in relation to the seemingly more specific array of people, events, and settings all of us encounter as we age[10]—is inadequately dealt with within extant organismic theories.

Because of the dissatisfaction with ideas derived from mechanism and organicism, much of the current scholarship in human development has gradually moved in the direction of developmental contextualism.[11] One way of understanding this change is to view developmental contextualism as integrating the active environment stressed in mechanism with the active organism emphasized in organicism. The active organism and the active environment are said to "interact" within developmental contextualism, but it is important to note that developmental contextual conceptions move beyond the simple notion of interaction found in organismic conceptions. Although nature and nurture may be separated conceptually for analytic reasons, they are functionally fused in life.[12] People do not exist outside of an environment, and the environment would have no meaning or importance if people were not in it, or at the least concerned about it.[13] In developmental contextualism, as I have illustrated in regard to the fusion between menarche and context, biology is seen as being shaped by the world within which the person exists, at the same time that the world is transformed by the biological nature of the organisms that populate it.[14] A young girl's family context will influence the way she reacts to the beginning of her menstrual cycle, whether she sees it as a positive or negative experience; and the broader context of life—involving historical changes in nutrition and medical care—can affect both the time at which menarche occurs and the attitudes of the girl and her family toward menarche.

It must be admitted that in developmental contextualism the relationship between nature and nurture is more complex than in mechanism or organicism. This complex relationship has been labeled dialectical,[15] transactional,[16] or dynamic interactional[17] and is the focus of the last two chapters of this book. Here, however, it is necessary to note that philosophical models about the nature and/or nurture basis of human development influence much more than scientific theory and research.

NATURE AND NURTURE IN SOCIAL LIFE

Most people, in all walks of life, have some conception of what human life is and how humans develop. Almost everyone has a conception of what it means to be human and from where the essence of that humanity derives, be it God, the social or cultural environment, or genes. Consequently, philosophical ideas about the basis of human development are not just products of and for scientists; they are attributes of the psychological functioning, of the ideologies, of virtually all members of a society.

The conceptions scientists have about the bases of human development often correspond, in their essential details, to those of the nonscientific public.[18] Such parallelism is not surprising. Scientists are part of the social and physical world they study. They may set forth their versions of these philosophical ideas in terms that differ from those their nonscientific or nonscholarly societal counterparts use, but at any moment in the history of a given culture there tend to be only a few distinct models of human development,[19] and therefore only a few distinct conceptions of the basis of human development.[20] These conceptions are shared by broad, crosscutting segments of the population. Poets, professors, and politicians alike may believe that a person's talents are God-given, passed on in the genes, or just shaped by the fortuitous confluence of social forces.

The specific content of these conceptions may vary within and across societies,[21] and because of their ubiquitous social presence these conceptions can have pervasive and penetrating effects on our lives. For instance, consider one society in which the basis of human development—of human talents, abilities, and attainments—is thought to derive largely from a person's genetic inheritance. And consider another society, wherein such developments are believed to be molded and promoted by the socialization and educational experiences within the particular culture.

In the first society, differences between people in their achievements are attributed to their differential breeding—that is, to their possession of different sets of genes. In the second society, different achievements must be attributed to contrasting histories of experience. But such differential views about what goes into making for success or failure across development may influence more than just how others in society interpret one's level of attainment across life. Beliefs about the basis of development can also influence government policies, programs, and laws.[22]

For instance, in the first society it might be argued that it is appropriate to enact laws for programs ensuring that the only people who should be allowed to reproduce are those who are free of genetic diseases that limit achievement. To give a simple example, society might not allow people with Down's syndrome to reproduce. If achievement in life is due to the possession of healthy genes, then it might make sense to argue that it is in society's interest to take steps to maximize the probability that people who have healthy genes will contribute to the future population. It might also be argued that people who do not possess such genes should be prevented from contributing much to the future population. People in this first society would believe that such a course of action would help promote the welfare of the state and the future health of its children.

In the second society, similar concerns for health and welfare would no doubt exist, but policies that address these concerns would probably not involve issues of genetics and breeding. Instead, the emphasis would be on enacting legislation to provide high-quality, stimulating, and enriching home, school, and community settings within which children would develop.

Indeed, the differences in the conceptions about the basis of human development that characterize our two hypothetical societies would influence more than laws and policies. Those conceptual differences would shape educational practices too, and consequently open or close doors to professional and vocational oppportunities. In a society of the first type, a legitimate question regarding educational policy and career development might be whether it makes good pedagogical and financial sense to offer people of all genetic backgrounds the same sets of training or vocational opportunities. Members of such a society might ask, "Are people, no matter what their race, ethnicity, or gender may be, equally capable of similar levels of learning in all subject areas?" It might also be asked whether people in these different groups are equally capable of playing similar roles in society. Given that these different groups differ in genetic makeup also, it would be argued, is it not best to offer educational experiences and provide social roles that fit the distinct capacities—which are shaped by the complement of genes—of the people in these different groups?

Different conceptions about the basis of human development can also influence general social intercourse, both within our families and in our communities. How do we treat those less fortunate than we? How do we act toward the poor or the homeless? How do we regard criminals or the mentally disturbed, and what do we believe about the possibility or even the desirability of attempting rehabilitation with them? How do we behave toward those who hold beliefs and engage in behavior we dislike or find reprehensible or believe is immoral? What do we tell our children about such people? How do we explain to our children, and to ourselves, why these people are different? Do we contend, for instance, that these people are inherently bad—that because of their temperament, constitution, or heredity they are intrinsically and inevitably the way they are? Or do we teach our children—through word and example—that these differences do not constitute differences in basic human value or in fitness in society, that the differences in how these people behave and in what they believe reflect only their distinct experiential history?

Finally, beliefs about the basis of our development affect our individual psychological functioning and behavior. People from our first society might believe it crucial to protect the health and reproductive ability of self, of family, and of other people of common heritage. People from our second society might believe it essential to provide oneself and others with certain experiences (e.g., educational ones) that are linked to the development of attributes of value and importance (e.g., intellectual attainments). However, people from *either* of these societies will act, as individuals and as members of a group of other like-minded members of society, to promote social policies believed to foster the attainments they value, be they matters of genetic health or of quality educational and social experiences. For instance, we might act to teach our children that the things we believe are essential to maximizing future development. In our first society people may be concerned with such topics as medical hygiene, genetics, or evolution. In the second society people may believe that curricula to promote exploration of complex situations and engagement in problem-solving activities must be developed. In either case, we will raise our children in hopes of inculcating in them similar commitments to those beliefs about development across life to which we adhere.

The actions we take to counter policies and people who stand in opposition to our cherished or core beliefs are as important as what we will act to promote. We may oppose policies we believe are destructive of humanity or diminishing the quality of human life and development; we may take exception to educational practices that attempt to imbue in our children values or beliefs we find objectionable or false; and we may not tolerate in our families or communities behaviors or people that destroy the core defining features of life we see as central to living lives of value and importance. In our first society, people may act to counteract those who would destroy—through illicit drugs, environmental pollution, or promiscuous or pornographic sexuality—the healthy reproductive capacities of our children. In our second society, people may act to oppose the same things, but for reasons linked to negative socialization experiences or to improper learning. Moreover, the same emphasis on nurture may make it more likely that members of the second society will oppose taking resources from public schools, or depriving minority children of quality day care, or restricting the open flow of ideas.

You and I may agree with some or all of the actions of members of either or both societies. We may see some or all of either set of actions as good (or bad). But such judgments are beside the point I am making here: that in both societies distinct social policies emerge because of differences in beliefs about whether nature or nurture—genes or environment—determine human development. Good or bad social policies may arise from either philosophy.[23] Nevertheless, some sort of implication inevitably arises from either philosophy, and in most cases these implications are quite different.

BIOLOGICAL DETERMINISM AND SOCIAL POLICY

Philosophical ideas about the basis of human development are powerful forces in our social worlds. They can profoundly influence social policy and social life, and that influence is a key focus of this book.

For several reasons, it is extremely important for citizens of contemporary societies to become informed about the influence of such ideas on social policy and social life. First, the two societies I have used as examples are not all that imaginary. The beliefs about the roles of nature or of nurture—of genes or environment—which were used to characterize the two societies, have existed in previous historical periods and indeed exist today.[24] The former type of belief, of nature over nurture, may be labeled as one of biological, or genetic, determinism. The second type of belief, of nurture over nature, may be termed social, or cultural, determinism. Either type of view is too extreme, in that both are scientifically inadequate.[25] Some critical data demonstrating this inadequacy will be discussed in this book. Despite the scientific limitations of both types of views, however, they continue to influence social policy and behavior.[26]

Their scientific inadequacies notwithstanding, these influences exist both within and outside of the scientific community, and even a cursory appraisal of the history of science will show that this continued influence is not unusual.[27] Because scientists are part of the social worlds they study, their ideas are not dissociated from the social world and consequently may subserve purposes that are not scientific. Indeed, the second reason my focus on the powerful influence of ideas about the basis of human development is important is that such ideas about nature and nurture have subserved and continue to subserve scientific, political, and social purposes.[28]

If the idea of environmental determinacy is scientifically unsound, it is because its picture of the ability of the environment to transform human development is too optimistic; humans are not so malleable that they can be thoroughly changed either organically or behaviorally.[29] In turn, if the idea of biological determinacy is scientifically unsound, it is because it depicts the nature of human life in terms that are too pessimistic; [30] humans are not automatons directed rigidly by a fixed genetic blueprint. And both views fail in that they do not integrate the processes from multiple levels of organization in order to study development.[31] (One level of organization would be biology, composed, for instance, of genes, hormones, and nerve cells. Another level would be the person, composed, for instance, of motivation, personality, and intelligence. And another level would be the environment, composed, for instance, of families, schools, communities, and the physical ecology.) Consequently, both views fail to take into account that at different stages of life and at each level of organization there are changes in the significance of the contributions of the different processes.[32]

Nevertheless, because ideas of both environmental determinacy and biological determinacy influence political and social life as well as science, the optimism or the pessimism they respectively involve will affect human development.[33] If the optimism of environmental determinacy is too great, it may lead to promises of improved social conditions and life attainments which cannot be kept. At the extreme, ideas associated with environmental determinacy can, in totalitarian regimes, be readily adopted to lend seeming scientific legitimacy to attempts to retrain or reorient people.[34] For instance, the victims of Stalin within the Soviet Union were relocated, worked to death, or simply killed, all in an attempt to change them through reeducation, based on a belief in "infinite plasticity" put forward by Soviet scientists, such as "Lysenko and other antigeneticists (ignorant of the real intellectual basis for environmentalism)."[35] Similarly inhumane and murderous reeducation programs based on the doctrine of cultural determinacy were part of the Red Guards' "Cultural Revolution" in the People's Republic of China in the 1960s. In Cambodia during the 1970s a program of deportation and mass extermination was based completely on a view of the extensive plasticity of human behavior; the human destruction that occurred has been estimated at about one-third that "accomplished" by the Nazis.[36]

In turn, the biological determinist's severe pessimism can deprive people of the opportunity to actualize what may be their very real but as yet undeveloped talents and abilities. At its most extreme, such pessimism may so unduly constrain people's initiative and action as to rob them of their rights, their humanity, and even their lives. Exploration of this view is the main goal of this book.

The doctrine of "nature over nurture" has been ubiquitous in human society at least since the time of Plato, who believed that a human's soul was innate and eternal, and thus unchangeable.[37] Similarly, persecutions in Spain from the fifteenth through the seventeenth century were based on church laws and civil statutes about the unchangeable purity or impurity of a person's blood.[38] Catholics whose families had converted from Judaism even centuries earlier were persecuted solely because of hereditary impurities, believed to be immutably present in bloodlines regardless of generations of church membership. And no less an institution than the American cavalry, in campaigns against Native Americans throughout the nineteenth century, took actions predicated on a similar biologically determinist conception, specifically the immutably savage nature of the "Indian population." As an illustration, consider the report of a Robert Bent, who in the mid-1860s was part of a cavalry action led by a Colonel Chavington:

> I saw the American flag waving and heard Black Kettle tell the Indians to stand around the flag. . . . I also saw a white flag raised. These

flags were . . . so conspicuous . . . they must have been seen. . . . I think there were six hundred Indians in all . . . thirty-five braves and some old men. . . . After the firing . . . I saw five squaws under a bank for shelter. When the troops came up to them they ran out and . . . begged for mercy, but the soldiers shot them all. . . . There seemed to be indiscriminate slaughter of men, women, and children. There were some thirty or forty squaws collected in a hole for protection; they sent out a little girl about six years old with a white flag on a stick; she had not proceeded but a few steps when she was shot and killed. All the squaws in that hole were afterwards killed, and four or five bucks outside. The squaws offered no resistance. Every one I saw dead was scalped. I saw one squaw cut open with an unborn child . . . lying by her side. . . . I saw the body of White Antelope with the privates cut off, and I heard a soldier say he was going to make a tobacco pouch out of them. I saw one squaw whose privates had been cut out. . . . I saw a little girl about five years of age who had been hid in the sand; two soldiers discovered her, drew their pistols and shot her, and they pulled her out of the sand by the arm. I saw quite a number of infants in arms killed with their mothers.[39]

The horror of such actions makes one wonder who the savages in the above account really were. Yet to Colonel Chavington the answer was clear. In a public speech he advocated the massacre and scalping of all Indians, even infants, by asserting simply, "Nits make lice!"[40] It was this very sentiment about the immutable uncivilized and less-than-human status of Native Americans that led U.S. General Philip Henry Sheridan to state, "The only good Indians I ever saw were dead," which as passed on became "The only good Indian is a dead Indian."[41]

Thus, for at least two thousand years of recorded history social actions have been implemented based on the belief that certain people had something inherent in them, something in their blood, that made them less than human and consequently deserving of persecution or even death. It was not until the mid-nineteenth century, however, that this doctrine became broadly legitimated in society and science.[42] Several intellectual events gave rise to this prominence. First and most important was Charles Darwin's view of evolution, involving the ideas of natural selection, survival of the fittest, and continuity in the biological heritage of animal and human—that is, of the descent of humans from prehuman ancestors.[43]

Second, this connection between humans and animals was extended to the social world, especially within Social Darwinism, a view that had its origin in the evolutionary thought of the English philosopher Herbert Spencer.[44] Indeed, Spencer coined the phrase "survival of the fittest" in reference to the evolution of cultures, although it was Darwin who adopted the term to describe the outcome of the process of natural selection.[45] The integration of Spencer's conception with Darwin's theory of the evolution of species

> produced a seemingly scientific rationalization of the 19th century European and American view of the peoples of the world as two populations, one of which was superior to the other by reason of physical and mental characteristics. . . . This rationalization came to be known as Social Darwinism. . . . [This view] arose during the most active period of industrialization and developing colonialism. The issue was the weeding out of the weak, the ill, the poor, the "socially unfit." . . . The "survival of the fittest" was an appropriate concept for that goal.[46]

A statement by John D. Rockefeller epitomizes the Social Darwinist perspective: "The growth of a large business is merely a survival of the fittest. . . . This is not an evil tendency in business. It is merely the working out of a law of nature."[47]

The outcome of the formulation of Social Darwinism was that humans were placed in the universe as biological, not spiritual, entities.[48] The view that emerged was that humans were molded gradually, over eons, to possess their essential, defining characteristics: intelligence, toughness, competitiveness, aggression, mastery. These attributes were the ones that, shaped by natural selection, enabled humans to survive the rigors and struggles of a hostile environment, a natural milieu where resources were limited, where other organisms competed fiercely for these resources, and where as a consequence only "the best" survived.

A cousin of Charles Darwin, the eminent English scientist Francis Galton, furthered this view by proposing in 1883 a new science—eugenics—aimed at ensuring that more of "the best" in fact survived. Galton coined the term "eugenics" from the Greek meaning "good in birth" or "noble in heredity" and intended his new science to give "the more suitable races or strains of blood a better chance of prevailing speedily over the less suitable."[49] Galton's "genius" not only launched this new approach to enhancing the survival of those noble of blood or heredity but also buttressed his work with mathematical and statistical analyses.

Indeed, increasingly since the later years of the nineteenth century, this view of the natural genesis of human abilities has been coupled with ideas from the precise and prestigious science of genetics.[50] As such, a material (or molecular) and a mechanistic facet have been added to this biologized view of human capacity: that the attributes that have been naturally selected over the course of human evolution are carried by genes. It is a human being's particular complement of genes (his or her genotype), given at conception, that provides the material basis of that person's evolved attributes.

GENETIC DETERMINISM: ONE STORY LINE

What, then, is the import of the merger of biological determinism and genetics for elaborating a line of reasoning about the implications of genetic inheritance for individual behavior and development? To help answer this question, suspend your disbelief for a few moments and consider one possible "story line" associated with the "doctrine of genetic determinism"— that is, with the translation of the general ideology of biological determinism into the more specific form of genetic determination.

From the perspective of genetic determinism, it is not what any of us does in life, not the environment we encounter, that is involved in the manifestation of our particular set of behaviors. Instead, those behaviors are functional outcomes of our biological heritage, our genes. Thus, what we do, what we become, is built into us at conception, is biologically predetermined. As a consequence, those who are the most able to compete, who have achieved and mastered the world, and who therefore occupy positions of power and prestige, have those places in society because of what has been inherited, not because of what has been encountered or learned.

Simply, then, according to genetic determinists, heroes and champions are born, not made. Our leaders, our "Best," are in the proper position in society because they have inherited the best genes, the genes that—as a consequence of survival over the eons of human evolution—are the most fit, the most adaptive. Those genes allow for maximum coping with the threats and challenges of a less than completely bountiful world. But of course not everyone is a leader, not everyone is in a position of power, not everyone is the Best. Those who occupy the lower social positions have not inherited the optimal set of genes. Thus, a person's place in the social hierarchy is genetically, not socially, determined. And therefore our roles in society match, or fit, the genes we possess.

If society is to meet and adapt to the challenges it will confront—indeed, if society is to survive—steps must be taken to ensure the maintenance of this social order. We have survived because the most adaptive genes—those shaped by natural selection to promote coping successfully with the world's challenges—have put our industrial, political, and military leaders into their deserved niches, and put the poor, the weak, and the powerless into their appropriate places in society. It would be counter to the laws of nature to destroy this hierarchy. Indeed, society must be vigilant to protect this social order if it is to continue to survive. Most important, society must be sure that the genes of the Best survive. In this view, certain artificial social conditions should not be allowed to alter the outcome of eons of natural selections. We must not allow society to introduce a "negative social selection," a societal situation wherein the genes of those who are not among the Best are permitted through social welfare, charity, or other similar programs either to be reproduced without restriction or, most problematically, to intermingle with and thus dilute the genotypes of the Best.

In sum, then, the doctrine of genetic determinism, when extended into the social realm, can provide a legitimating platform for a political philosophy that upholds the status quo as biological imperative and indeed as moral necessity. Although this political extension is not a necessary derivative of the doctrine of genetic determinism, when people do think in this way they are committing the "naturalistic fallacy"—they are translating a particular social situation, something that "is," into something that is morally required, something that "ought" to be. Moreover, such political adherents are able to appeal to "scientific" evidence in defense of their cause. The doctrine of genetic determinism offers them a scientific basis for justifying the established social hierarchy according to biologically determined principles. This, in turn, can lead to the placing of differential value on people: the powerful, the prestigious, are more valuable to society, and those lower in the hierarchy are less valuable. The powerful and prestigious are the people who will ensure the future health and prosperity of the nation—in effect, societal survival—but if the genes of those lower in the social hierarchy become too prominent, society will be endangered. At the extreme, if the nation faced a potentially calamitous threat to its survival, reliance on the inferior genes would be suicidal.

Because carriers of "inferior" genes can hardly be seen as superior or desirable people, a prejudice both against the genotypes and against the people possessing them readily arises: these people simply cannot do as much for society, and moreover they should not be allowed even to try, because of inevitable societal deterioration that would follow. Furthermore, in this view, it makes good scientific and societal sense to take steps to protect the Best, to ensure the healthy reproduction of their genotypes. How might this be done? Perhaps there should be laws restricting marital unions between those with positively valued complements of genes and those with negatively appraised genotypes—for instance, antimiscegenation laws. Perhaps there also should be reproduction laws curtailing the potential of negatively evaluated people to contribute their genes to the future population of society. Sterilization laws are one such possibility.

And perhaps—so goes our scenario—when even the Best are being tested at their limits, in times of dire national emergency or when the survival of the society is at stake, more draconian measures should be taken to ensure that the Best survive. In such dramatic circumstances, when resources for the survival of the Best are so severely constrained as to bring even subsistence into question, a sort of "social triage" must be undertaken. Quite special treatment—*Sonderbehandlung*, to use the Nazi euphemism for the gassing and cremation of millions of carriers of the undesirable genotypes—might be seen as justified. The problem those undesirable genes represent—of both diluting and polluting the genetic capacity of the Best people—must be solved in some all-encompassing manner if this acute threat to the survival of the society is to be met. Some definitive solution, some final solution, must be attained in order to ensure that the degenerating genes will not mortally wound the genetic health of the Best in society.

It follows that the "special treatment" needed would be destruction of those threatening genotypes. If that means the carriers of those genotypes must also be destroyed, as it obviously does, then there is really no choice, however difficult or regrettable this action may be. Indeed, not to take such action would be an immoral act toward one's own, to others in the community shared by the Best.

And finally, so the argument goes, do not carriers of the inferior genes deserve the special treatment they receive? Are they not worthy of our enmity and our hate? After all, it is they who diminish and threaten the survival of all that is valued in society; it is they who, if allowed to flourish, would destroy the adaptive outcome of eons of evolution. In fact, are we not ultimately doing these people a favor, saving them from a fate of living as less-than-human, mongrelized beasts in a degenerate world destined for certain destruction for lack of genes that would protect the life of the society?

GENETIC DETERMINISM YESTERDAY AND TODAY

Is this journey I have taken you on—from the pessimism of the doctrine of genetic determination, through the doorway of what may be seen as social class, ethnic, and racial prejudice, down the hallway of social, institutional, and governmental discrimination, and to the threshold of genocide—a flight of fancy? Unfortunately, the answer is no. The same story line of genetic determination can be seen at least in part again and again across even our recent history: in the U.S. Cavalry's military campaigns against the Indians;[51] in the treatment of African-Americans by European Americans in the United States; in the annihilation of Armenians by the Ottoman conquerors in the early part of the twentieth century;[52] in the continuing apartheid policies in South Africa; in the rise of the fascist "National Front" in contemporary Great Britain;[53] and in the Holocaust that engulfed European Jewry with the avalanche of National Socialism, Nazism, in Germany from 1933 to 1945. In the case of the Nazis, all facets of the above story line were actualized in the societal enactment of the doctrine of genetic determinism in the most stark, grotesque, and abhorrent manners, involving antimiscegenation laws; compulsory sterilization; forced euthanasia (to employ a bizarre oxymoron) of mental patients, the aged, the infirm, and the politically, ethnically, and racially undesirable; and finally the mass murders, the gassing and the cremation, of millions.[54] Such was the "special treatment" the Nazis believed was necessary to rid their world of the threat of degenerate genes.

In Nazi Germany we see, then, the clearest instance of the *coupling* of the doctrine of biological determinism with a political movement based on racism. This coupling of "science" and politics gives the appearance of intellectual legitimation and of social urgency to the policies called for by such a political movement. These political movements build the rationale for their existence on the basis of the threat to those "in" the movement by those "out" of it. In other words, these political organizations coalesce around the belief that the world, or at least their sphere of influence, may be divided unequivocally into two major groups: an "in-group" comprising those possessing the best human characteristics, and an "out-group" comprising those possessing the worst features of human existence. There can be no crossing-over between these groups, because blood, or genes, divides them. Thus, the doctrine of biological determinism exists ready for co-optation by proponents of such a political movement. It provides a compelling rationale for the basis of the immutable differences between the in-group and the out-group. By couching these differences in biological, or scientific, terms, the "truth" of the in-group's racist views is seemingly proven as an empirical fact of science. Moreover, casting the differences between groups as biological differences makes the threat the out-group poses not one of mere political advantage, but rather one of absolute survival. The in-group, when in control of political power, must exercise this influence completely, and take even the most draconian of measures, in order to protect against the danger of biological annihilation posed by the out-group.

It is crucial to understand that the connection between the doctrine of biological determinism (or, more specifically, of genetic determinism), on the one hand, and racism and political movements, on the other hand, is neither a necessary nor an isomorphic one. Biological determinists are not automatically racists or proponents of political movements based on biological determinism. In fact, many scientists who espouse ideas consistent with biological determinism adhere to positions that are politically quite liberal or egalitarian.[55] Therefore, biological determinists are not necessarily racist or politically conservative, and certainly, then, biological determinists are not automatically adherents of Nazi ideology or any corresponding fascist political beliefs.

However, Nazis (or, more generally, fascists) *are* necessarily racist, and *in addition* adopt the ideology of biological determinism in order to legitimize the racist cornerstone of their political philosophy.[56] In other words, the social-policy implications of the doctrine of biological, or genetic, determinism arise when the ideas of this doctrine are *coupled with* fascist political philosophy. Simply, genetic determinism is co-opted to serve the political agenda of fascism. If such a biological ideology did not exist ready for such exploitation, fascists would have to invent it.

Michael Billig's analysis of the components of fascist movements makes a complementary point. Billig believes that the first and key defining feature of fascist political programs is a nationalism built on racism, that although racists need not be fascists, fascists must be racists because "above all, fascist ideology bases itself upon a strong belief in the unity of a particular race or nation; as such it will attempt to address its message to the whole of the nation/race, thereby seeking mass support."[57]

Biological determinism is the glue that holds the nation/race together and separates it from peoples of different nations or races—that is, those with other genes or "blood." Fixed and immutable biological differences separate members of the in-group (nation or race) from members of the out-group. All other defining features of fascism derive from this immutable biological difference and the threat to the in-group's biological purity and survival posed by the out-group. For instance, the well-known fascist antipathy toward Marxism and communism[58] is related to the former's emphasis on intranation or intrarace conflict (i.e., class struggles). Similarly, the Nazi persecution of Catholics (despite Hitler's being both born and raised a Catholic)[59] may be interpreted as having to do with Catholics' adherence to laws outside the German state—the laws of the church, through the Pope and the Vatican. Similarly, the fascists' abhorrence of democratic rule and personal freedoms[60] derives from the view that the state must be in total control in order to protect the in-group from the constant and insidious threat posed by the out-group.

For instance, writing in Great Britain in the late 1970s, Billig saw a rise in the political influence of a version of fascism quite reminiscent of German National Socialism. Akin to the Nazis, members of the British "National Front" have based their increasingly successful calls for a fascist government on a warning about the dangers of biological (genetic) degeneration posed by the presence of Blacks and Jews in British society. Billig contends that, when fascist political philosophy and biological determinist ideology are coupled in a self-confirming, tautological manner, an argument can be made for a pervasive, indeed society-wide, curtailing of personal freedom and democracy. He indicates that such restrictions can involve the institution of social policies ranging from discrimination to genocide. The self-reinforcing character of the merger between biological determinist ideology and political conservatism is based on the belief that immutable biological dangers posed by the biological character of an out-group exist for an in-group, that therefore political mechanisms must be organized to reflect and remedy these biological circumstances,[61] and that only through such policies can the genetic integrity of the in-group be protected.

Billig cites passages from the main political journal of the National Front, the *Spearhead,* to illustrate this relatively recent British manifestation of the merger of biological determinist ideology and fascist political philosophy. For example, *Spearhead* editorials describe Jews as "a race of unheroic, greasy, shifty-eyed, sickly money lenders, rent-racketeers, pornographers and big business wide-boys." Those British who are not infected by "the inferior genes from the Third World" are cautioned that "it is vital that every man in the street become Jew-wise. . . . If Britain were to become Jew-clean she would have no nigger neighbors to worry about. . . . It is the Jews who are our misfortune."[62]

These "editorial" comments may be readily dismissed as viscerally repulsive and mindless racist prattle, but the statements are markedly redolent of the type of Nazi propaganda that eventually inflamed the German citizenry to a level sufficient for them to give at least their acquiescence to National Socialist policies excluding Jews and other "genetic misfits" from society.

Thus, neither the doctrine of genetic determination nor its co-optation by fascist political ideologies died with the National Socialist party. In fact, biological and social science of today is faced with yet another version of the doctrine of biological determinism, one termed "sociobiology," which also provides fertile ground for co-optation by racist and/or fascist political movements. This version of genetic determinism has been heralded by its founder, Harvard biologist E. O. Wilson, as the "new synthesis," as the science of "the systematic study of the biological basis of all social behavior."[63] As in other versions of the doctrine of genetic determinism, sociobiology purports to provide a scientific legitimation of the established social hierarchy, a genetic justification of the different social roles played if not by Jews and Aryans then by men and women.[64] To give one example, the University of Chicago psychologist and sociobiologist Daniel G. Freedman asserts:

> As far as I can see, the male sense of omnipotence is part of an evolutionary heritage among hierarchically arranged species. . . . It is by now a well-known fact that women's groups *must* exclude men if the average woman participant is to speak openly. The very presence of men, however silent they remain, is inhibiting, especially to younger women. It can be described as a sort of reflexive "insignificant little me" response.[65]

The general theoretical terms contemporary sociobiologists use to explain their ideas of nature over nurture parallel quite dramatically those found in the writings of the scientists the Nazis relied on to promote their ideas of

Social Darwinism, of the genetic determination of behavior, and of the need for social policies ensuring the genetic health or racial hygiene of the German people. Moreover, the implications of genetic determinist ideas drawn by contemporary sociobiologists about sex differences in social behaviors and about the biologically "proper" role of women and men in society are quite similar to those of these Nazi theorists, who, it should be noted, wrote not only about the "Jewish problem" but also about "the woman problem."[66]

Contemporary sociobiologists are certainly not neo-Nazis. They do not in any way advocate genocide and may not even espouse conservative political views. Nevertheless, the correspondence between their ideas (especially regarding women) and those of the Nazi theorists is more than striking, and both sets of ideas are equally scientifically flawed. Moreover, as I have explained already, across history such ideas have been all too easily coupled with political programs based on racial prejudice. As the philosopher George Santayana warned, those who are unaware of history are condemned to repeat it. Therefore, the parallelism between Nazi genetic determinist ideology and sociobiological genetic theory must be of great and immediate concern, both because of the weak scientific grounds on which they are based and because of the pernicious (and scientifically ill-based) social extensions that can be derived from these ideas. We must avoid the propagation of social policies that, on the premise of legitimation by contemporary scientific understanding of evolution and genetics, unfairly discriminate between men and women. Such policies would again open the door to discrimination against other groups—racial, ethnic, or national—that can purportedly be differentiated on the basis of their genes.

The flaws in contemporary sociobiological thinking about genetic determination in general and about sex differences in social behavior in particular are akin to the intellectual errors found in National Socialist racial ideology. Indeed, both these sets of ideas exemplify genetic determinist thinking. When such ideas have been coupled with political regimes, as occurred in Nazi Germany, they have been associated with the delimitation, degradation, and destruction of millions of humans. Because the relevance of such thinking to our lives and to future generations cannot be overestimated, it is necessary to explain these genetic determinist ideas, to indicate their shortcomings, and to provide a scientifically defensible alternative to them. The rest of this book is my attempt to do just that.

2

A PATH TO MASS MURDER:
FOUNDATIONS OF
THE NAZI PROGRAM OF GENOCIDE

National Socialism is nothing but applied biology.
—Rudolf Hess, 1934

In modern Western industrial societies, scientists who pursue either basic or applied research can find financial support for their endeavors. Whether it is easier to obtain funding for basic or applied work, or whether one of these two research emphases is more or less valued, depends on the social needs and political currents prevalent in a society at a given time. It is extraordinarily rare, however, for an entire society to have as the primary aim of its governmental, political, legal, educational, and medical institutions the application of a given approach to a specific area of science. Yet, this broad subservience of an entire society to a particular view of biological science is precisely what occurred in Germany under the hegemony of the Nazis.

BIOLOGICAL DETERMINIST THINKING IN GERMANY

As the quotation at the beginning of this chapter indicates, leaders of the Nazi party, such as Deputy Führer Rudolf Hess, believed Nazism to be merely "applied racial science."[1] This conception—of biology representing the cornerstone of the state, the foundation of the Nazi view of the world—was echoed by Nazi politicians and scientists alike.[2] The core mission of the Nazis can be said to have been to apply a view of the genetic determination of the superiority of the Nordic, or Aryan, race to the peoples of Europe, if not the entire world. This application required a unification and subordination, a *Gleichschaltung,* within and across institutions in the society in order to coordinate their contributions to accomplishing this application.[3] The result would be a "primacy of national biology over national economy," a unified national effort "to put into effect the laws of life, which are biological laws."[4]

In essence, epitomizing a fascist state, the Nazis co-opted the view of biological determination of racial superiority in order to legitimate the unification of all facets of German life under their total political control. Biology and politics became completely merged. Indeed, Hitler saw the link between biology and a nation's policies as involving a part-whole relationship.[5] He contended that human nature is part of nature and therefore follows the same laws as the rest of nature. He believed that the laws of nature demanded inequality, hierarchy, and subordination of inferior forms to superior ones but that human history was a series of revolts against these demands of nature. Such revolts, in destroying inequalities between the inferior and the superior, led to ever greater egalitarianism and thus threatened the biological integrity (the survival value) of the superior people of a nation. As a consequence of this revolt against the natural biological order, Hitler believed, "there will be but two possibilities: either the world will be governed by the ideas of our modern democracy, and then the weight of any decision will result in favor of the numerically stronger races, or the world will be dominated in accordance with the laws of the natural order of force, and then it is the peoples of brutal will who will conquer." He said, in other words, "a stronger race will drive out the weak, for the vital urge in its ultimate form will, time and time again, burst all the absurd fetters of the so-called humanity of individuals, in order to replace it by the humanity of Nature which destroys the weak to give his place to the strong."[6]

But what was this biologically stronger race Hitler spoke of? Addressing this question provides insight into the cornerstone of the National Socialist biological determinist vision of the world. The Nazis' biological world view was based on one core idea: The German *Volk* (people) were, across evolution, naturally selected to possess the genes for leadership in and mastery over the world. To understand the full implication of this world view, one must understand that the Nazi concept of *Volk* connotes much more than the term "people." It refers to:

> the union of a group of people with a transcendental "essence" . . .
> [which] might be called "nature" or "cosmos" or "mythos," but in

each instance . . . was fused to man's innermost nature, and represented the source of his creativity, his depth of feeling, his individuality, and his unity with other members of the *Volk*. Here we may say
that *Volk* came to embody an immortalizing connection with eternal
racial and cultural substances.[7]

This mystical or transcendental vision of the genetic superiority of the
German *Volk* was not born in the mind of Adolf Hitler or in the propaganda
of Joseph Goebbels. Quite the contrary, the Nazis adopted a tradition in
German biological and medical science which placed within the genes of the
German *Volk* the basis of the salvation of humanity, a salvation made
necessary because of the threat of genetic pollution posed by miscegenation
with genes from subhuman or nonhuman populations.

THE ROLE OF ERNST HAECKEL

Much of the biological determinist thinking that would find its most
politically powerful and perverse actualization in Nazi Germany has its
roots in late nineteenth-century and early twentieth-century German Social
Darwinism.[8] The role of Ernst Haeckel (1834–1919), famed biologist,
Darwinist, and theoretician, is central.[9] Haeckel's work was a major intellectual force in bringing Darwin's work into German scholarship and in
creating the German Social Darwinist movement. Yet Haeckel's goal was to
provide scientific legitimation for the romantic vision of the German *Volk*
as a people who have singularly met the test of succeeding in the struggle
(*Kampf*) for survival imposed by nature and, as a consequence, of having
been selected for hegemony over other races, and indeed the world.[10]

In this synthesis of his biologically determinist version of Darwinian
evolutional principles and *volkish** philosophy, Haeckel forged a viewpoint
that was as much a political movement as it was science. Indeed, his work is
one of the prime historical examples of the commingling of biological
determinist ideas with racist political views, and as such it could provide an
obvious tool for exploitation by an emerging fascist state built on racist

---

*The proper German spelling of *volkish* is *völkisch*. The spelling *volkish* used in this book
is an adaptation common in English-language works. The meaning of the term remains the
same.

doctrines. Indeed, Haeckel's perspective was a forerunner of the National Socialist vision of the synthesis of biology and the policies of the fascist state. According to historian George Stein, Haeckel's views

> combined an almost mystical, religious belief in the forces of nature (i.e., natural selection as the fundamental law of life) with a literal, and not analogical, transfer of the laws of biology to the social and political arena. It was, in essence, a romantic folkism synthesized with scientific evolutionism. It included the standard Darwinian ideas of struggle (*Kampf*) and competition as the foundation for natural, and therefore social law, with a curious "religion" of nature which implied a small place for rationalism, the lack of free will, and happiness as submission to the eternal laws of nature. *Blut und Boden* were the reality of human existence.[11]

In 1906, joined by several prominent German scientists, theologians, literary critics, novelists, and politicians, Haeckel formed the Monist League, the aim of which was to organize both scientific and political support for Haeckel's Social Darwinist ideas.[12] The belief uniting the members of the Monist League was that all of life—human and nonhuman—could be unified through use of Haeckel's Social Darwinist principles. This one set of ideas could integrate not only the understanding of human evolution but politics, religion, morality, and ethics as well.

Members of the Monist League promoted, then, what is termed in philosophy the "naturalistic fallacy." This fallacy occurs when one argues that by observing what "is the case" (for example, by seeing the rules that do exist in a given society) one can know what "should, or ought to be, the case" (in regard to the rules of a society). In other words, the world we live in provides us with what *is* the character of our existence. In turn, our morality, or ethics, provides us with beliefs about what *ought* to be the character of our lives. Just because certain laws, social relationships, and customs exist in a given society does not make them just; their mere existence does not mean they ought to exist. However, the naturalistic fallacy is the translation of what *is* to what *ought* to be. For instance, if one sees (or claims to see) that a particular group (e.g., the German *Volk*) has struggled, through evolution, to ascend to its leadership position among all other "races of humans," this group ought to be afforded all advantages in society and take its deserved place in society. Indeed, according to this view that merges biology and politics, it is morally correct for all society's laws and political institutions to be shaped to subserve this purpose. Thus, the members of the Monist League constructed their ideology on the foundation of this flawed logic.

The goal, then, of the Monist League was nothing short of biosocial reform. Haeckel wanted his views of struggle, and of what he saw as the resulting qualities of the German *Volk,* to be regarded as epitomizing the ideal outcome of evolution; and the merger of biology, politics, and morality meant that this evolutionary "result" should become the building blocks of German political and social policies. Given this goal, and the prominence of both Haeckel's work and that of his colleagues in the Monist League, a striking resemblance exists between Haeckel's view of *volkish Kampf* and what we have seen to be Hitler's conception, expressed in his *Mein Kampf.* Hitler captured the "spirit of his times" (the *Zeitgeist*), which stressed German biological superiority through evolutionary struggle, in his own political persona. He transformed *volkish Kampf* into his own personal struggle and in so doing became the almost deified symbol of the entire German nation/race. In short, Haeckel's view of *volkish* struggle blended into Hitler's and as such into that of National Socialist Germany.

The correspondence between Haeckel and Hitler did not exist only at a theoretical level. Haeckel's belief in the superiority of the German *Volk* was coupled with his assertion of the inferiority—indeed, the subhuman charac-ter—of other peoples.[13] Given the adherence to the naturalistic fallacy, and the consequent merger of biology and morality, the inferior character of those people meant that their lives were of less value. The value of the in-group was all the more apparent when seen in contrast to the inferiority of the out-group.

To illustrate, Haeckel contended: "The morphological differences between two generally recognized species—for example, sheep and goats—are much less important than those . . . between a Hottentot and a man of the Teutonic [i.e., German *volkish*] race." It is the German *Volk* that has "deviated furthest from the common form of ape-like men [and that] outstrip[s] all others in the career of civilization." In turn, races of "woolly-haired [are] incapable of a true inner culture or of a higher mental devel-opment. . . . No woolly-haired nation has ever had an important history." Thus, Haeckel concluded, because "the lower races—such as the Veddahs or Australian Negroes—are psychologically nearer to the mammals—apes and dogs—than to civilized Europeans, we must, therefore, assign a totally different value to their lives."[14]

Given the lesser value of inferior races, as well as of people whose biological characteristics made them less than perfect physical representatives of the *volkish* ideal, Haeckel found it reasonable to suggest policies that would ensure the maintenance of the high evolutionary status of the German people. His suggestion was based on the view that if one allowed inferior genetic "stock" to be kept safe from the rigors of natural selection, and thus to be kept alive "artificially" by society (to be "domesticated"), then an overall degeneration of the biological (genetic) health of the *Volk* would result. This concept of "domestication-induced degeneracy," a degeneracy induced by society's termination of natural selection (through misguided humanitarianism, liberalism, welfarism, democracy, etc.), occurs again and again in Nazi and proto-Nazi ideological arguments (see the discussion on Konrad Lorenz in Chapter 3). The key basis of the domestication-induced degeneracy idea, however, was found in the thinking of Haeckel, who believed that to maintain the superiority and purity of the German *Volk,* social policies *ridding* society of inferior human "stock" would have to be instituted.

Haeckel recommended a "negative eugenics" strategy, whereby the racial purity of the in-group was improved not by selective breeding (a "positive eugenics" strategy)[15] but by terminating the lives of out-group members. Thus, Haeckel called for destruction of the "useless" lives of these evolutionary inferiors.[16] This appeal would be taken up a few decades later by one of the founders of the German eugenics movement, Wilhelm Schallmayer, who was also in Haeckel's Monist League, and it would become part of the explicit policies in Hitler's Third Reich.

To illustrate Haeckel's views, we can note his beliefs that infanticide of deformed and sickly newborns was "a practice of advantage to both the infants destroyed and to the community"; that the "utterly useless . . . [the] hundreds of thousands of incurables—lunatics, lepers, people with cancer, etc.—who are artificially kept alive . . . without the slightest profit to themselves or the general body" should similarly be killed to attain their "redemption from evil"; and that capital punishment "for incorrigible and degraded criminals is not only just, but also a benefit to the better portion of mankind [since by] the indiscriminate destruction of all incorrigible criminals, not only would the struggle for life among the better portion of mankind be made easier, but also an advantageous artificial process of selection would be set into practice, since the possibility of transmitting their injurious qualities would be taken from those degenerate outcasts."[17]

Thus, in the writings of Ernst Haeckel and his associates in the Monist League, the theoretical link between biology and state policy which would be adopted by the Nazis was established by the beginning of the twentieth century. In addition, the prescriptions for state-sponsored murder, performed for the purported purpose of protecting the state from biological annihilation by subhuman degenerates, would become the enacted policies of National Socialist Germany. Indeed, George Stein concludes his analysis of ethnocentrism, racism, and nationalism in Nazi Germany by noting: "There really was very little left for national socialism to invent. The foundations of a biopolicy of ethnocentrism, racism, and xenophobic nationalism had already been established within German life and culture by many of the leading scientists of Germany well before the First World War."[18]

Haeckel's arguments for the biological (genetic) determination of physical and behavioral differences among the races and his contention that the major "races of mankind" (e.g., the "Caucasoids," "Negroids," and "Mongoloids") were in fact separate species[19] paved the way for some of the specific arguments Hitler would later make (for instance, regarding Jews as an antirace).[20] Although the Nazis did not pay as much attention to differences between blacks and whites as they did to differences between Aryans and Jews, the fact that Haeckel made equally strong anti-Semitic statements surely helped establish his influence among Nazi biological determinist ideologues.[21]

ROOTS OF ANTI-SEMITISM

"Anti-Semitism," a term coined in Germany in 1879 (*Antisemitismus*),[22] was part of the biologized view of the social world that permeated late nineteenth-century Germany. This integration was found primarily in the writings of Paul de Lagarde (1827–91), whom some historians regard as the patron saint of the German anti-Semitism movement.[23] It was Lagarde, a prominent professor of Asian studies and writer on religious issues, who popularized the idea that the biological nature of the Jew was threatening the *volkish* community, or *Volksgemeinschaft*.[24] Calling for the transformation of Christianity into a Christian-German religion, Lagarde sought to eliminate all Jewish elements from Christian life.[25] The urgency of this need for extermination was stressed because, to Lagarde, the Jew was inherently a carrier of biological disease. To represent Jews, Lagarde used concepts and terms that, we shall see, correspond exactly to those used by Hitler and other Nazi ideologues, as well as by Lorenz. Lagarde believed the Jewish threat to the German *Volk* was equivalent to the threat of infestation by infected vermin or by diseased bacilli:

> One would need a heart as hard as crocodile hide not to feel sorry for the poor exploited Germans and—which is identical—not to hate the Jews and despise those who—out of humanity!—defend these Jews or who are too cowardly to trample this usurious vermin to death. With trichinae and bacilli one does not negotiate, nor are trichinae and bacilli to be educated; they are exterminated as quickly and thoroughly as possible.[26]

These ideas about innate racial differences and about the threat to the health of the German race represented by the spread of diseased, "inferior," races were soon to be integrated by the promulgation of a "new" German biomedical vision. That conception, it was believed, would bring health and purity to the German race by introducing medical procedures and social policies aimed at eliminating the possibility that diseased and/or biologically degenerating races would be allowed to threaten the *Volk*.

THE GERMAN RACIAL HYGIENE MOVEMENT

In 1895 a Social Darwinist, Alfred Ploetz, published *The Excellence of Our Race and the Protection of the Weak*, which crystallized, within Germany, genetic determinist ideas about the evolutionary superiority of the German *Volk*.[27] Consistent with Haeckel's vision of the synthesis among biology, politics, and moral worth, Ploetz believed that "race was the criterion of value" in society and that because of the superordinate value of race "the State is not there to see that the individual gets his rights, but to serve the race."[28] However, in the German state as Ploetz viewed it, existing social policies were interfering with the advancement of the race: society had instituted programs that amounted to the "negative social selections" I have already mentioned. Raising again the Haeckel-like argument of domestication-induced degeneracy caused by societal termination of the forces of natural selection, Ploetz indicated that misguided humanitarian action was endangering the quality of the race by allowing the protection, survival, and reproduction of weaker members,[29] and that such events as wars and revolutions had indeed led to the demise of the best of the race—the strong, heroic, and combative young men—while the weak and the sick, who did not participate in battle, survived.[30] Such differential survival rates of the best and the worst genotypes, respectively, resulted in racial degeneration.

Ploetz accordingly called for biological or medical intervention to address the problems posed by poor people producing too many children. In order to redress the problem of impending racial degeneration, he recommended a reversed view of medical care: Do not make providing medical care for the individual the primary goal of the physician, for such an emphasis may threaten the race; instead, the physicians should make the health of the entire race their primary concern.[31] Ploetz termed this new approach to medicine "racial hygiene" (*Rassenhygiene*) and suggested delaying support for the poor until they are past childbearing age, and withholding medical care from the weak (because such intervention would promote the survival and reproduction of genes that would otherwise have died).[32] Perhaps most significant, Ploetz proposed new forms of human breeding: because both the negative and the positive characteristics of the race were genetically based, only genes promoting positive attributes should be allowed to reproduce.[33] In other words, Ploetz called for replacing the negative social selections currently at work in society, those permitting the weak and the misfits to survive and reproduce, with what would be (in his view) a positive social selection. Since human beings had destroyed the natural selection process through misguided social programs, a new social program must put humans back on the correct path, the path of preserving and promoting the propagation of the genes of the best of the race.

The racial hygiene vision of Ploetz captured the minds of German Social Darwinists, who also believed that what they regarded as the liberal and humanitarian social policies existing in Germany were destroying the process of natural selection and that there was a danger that the poorly fit would reproduce more rapidly than the fit.[34] Ploetz's call for a new breeding program to preserve the race, related as it was to the *Zeitgeist* that Haeckel and his Monist League helped foster, became a mission that others adopted. For instance, in 1903 Wilhelm Schallmayer, a member of the Monist League and a founder of the German eugenics movement, argued that because misguided social institutions were saving the weak and the sick, natural selection could no longer be relied on to perfect the race. He called for social selections designed to guide the race in the "correct" direction—that is, used to disallow the survival of people who would die without the aid of social institutions.[35]

Schallmayer appealed for society to institute biomedical procedures to gain control over which genes would survive and which would not. The controls for which he called involved withholding aid to and/or medical care for the ill, the weak, and the poor. But although these procedures diminish the groups' chances for survival, they do not involve the direct selection of people for murder. Yet Haeckel called for just such action. His scientific prominence and that of his associates in the Monist League apparently furthered a social climate suggesting to others the utility of such actions. Thus, thirteen years prior to Hitler's assumption of power, and in fact four years before the publication of his *Mein Kampf*, selection based on the killing of purported inferiors was explicitly called for by Karl Binding and Alfred Hoche in their 1920 book *Die Freigabe der Vernichtung lebensunwerten Lebens* (The Sanctioning of the Destruction of Lives Unworthy to Be Lived).

Binding, a professor of jurisprudence at the University of Leipzig, and Hoche, a professor of psychiatry and director of the psychiatric clinic at the University of Freiburg, presented their solution to a problem identified also by Ploetz. Writing shortly after the end of World War I, they pointed out that the best young men died in war, which meant that the *Volk* lost the best available genes. The genes that proliferated were those of people who did not fight, genes that were by definition, through the fact that their possessors did not participate in the war, inferior. Such a process increased both biological and cultural degeneration,[36] which, given the reduction of all social, moral, and political life to genes, were in any case interchangeable. That situation was intolerable, raising again the specter of domestication-induced degeneracy, and it led Binding and Hoche to combine their legal and medical expertise to argue that the state should direct the killing of "worthless" people. Among those appropriate for such killing were individuals who were "mentally completely dead" and people who "represent a foreign body in human society." Both groups are among those "who cannot be rescued and whose death is urgently necessary" to protect and enhance the quality of the race.[37]

These examples of the history of genetic determinist thinking in Germany prior to the advent of the Nazi era—involving the Social Darwinist Monist League, the eugenics movement, and the "new science" of racial hygiene—indicate that neither racist doctrines of National Socialism nor the ideas for their application to the social world either began with Hitler and his cohorts or were concentrated in a small group of socially marginal fanatics.[38]

Spurred on by similar hereditarian (eugenics) movements in Great Britain and the United States,[39] the genetic determinism of these German movements was a doctrine permeating German biological and medical science. The biologizing of prejudice, discrimination, and ultimately the call for genocide was invented and promoted by "normal scientists," and indeed by leaders within their professions.[40] Not only can these scientists, in hindsight, be regarded as among the top professionals in their fields at the time of their work; they also saw themselves with some justification as having the same scientific status as such people as Pasteur, Koch, and Lister.[41]

Thus, the biological determinist ideas that crystallized within the German racial hygiene movement were both supported and furthered by people at the most prestigious, honored, and influential levels of German biomedical science, such as Eugen Fischer, Ernst Rudin, Otmar von Verschuer, Rudolf Ramm, and Gerhard Wagner.[42] So National Socialism did not introduce into German society a *new* genetic determinist ideology. Quite to the contrary, once the Nazis assumed power the statewide propagation of racist ideology and its enactment into actual social policies were only furthered.

FROM IDEOLOGY TO POLICY: BIOLOGICAL DETERMINISM DURING THE THIRD REICH

The heritage of Ploetz and after him of Schallmayer, of Binding and Hoche, and ultimately of Haeckel and the Monist League was extended by Germany's first professor of racial hygiene, Fritz Lenz, who was appointed to his position at the University of Munich in 1923. Due to the prominence of his appointment, and to his co-authorship (with Eugen Fischer and Erwin Baur) of perhaps the most important German textbook on genetics and racial hygiene, Lenz had a broad influence on the thinking of scientific colleagues and students as well as on politicians who believed, as Rudolf Hess claimed, that National Socialism was only applied biology. Lenz's influence was decidedly one of furthering the ideology of genetic determinism.[43]

Lenz's views about the genetic determination of racial superiority may be seen in his answer to several questions he posed to himself: "Why is it that many persons are able, many others stupid, and the majority mediocre? Why are some people cheerful and others gloomy; some industrious and others slothful; some unselfish and others selfish?" Lenz answers: "Characteristics of the mind, no less than those of the body, are rooted in the human hereditary equipment. . . . Environmental influences (including education in the narrower sense of the term) can do nothing more than help or hinder the flowering of hereditary potentialities."[44] Thus, for Lenz the primary basis of human cognitive, personality, and social behavior lies in the genes, with environment playing only a secondary and certainly noncausal role.

In essence, then, Lenz contended that individual differences in human characteristics, encompassing a set of behaviors that can be characterized as ranging from successful to unsuccessful, from best to worst, or simply from good to bad, derive from one's genetic heritage. In short, Lenz saw a correspondence between genetic differences and moral ones, as did Haeckel. Moreover, in both conceptions environmental influences play no role in causing or shaping the essential nature of one's character.

Given these views, the features of the racial hygiene movement, as Lenz conceived of it, are clear. First, it is obvious that the power of the environment to shape human behavior is dismissed. In fact, Lenz contended that the belief that environment could influence either individual behavior or social institutions was merely a political, ideological idea, not a scientifically based viewpoint. Second, Lenz contended that the "decisive motivating force in racial hygiene" was the idea that "not all evil is determined by the environment, and that the roots of most evil lie instead in hereditary defects."[45] This conception thus connects the genetic determination of behavior to a moral valuation of the genes producing the behavior: some genes produce evil behavior, some good behavior. Given the isomorphic relation between genes and behavior, it follows that there are both evil and good genotypes, carried by evil-behaving and good-behaving people, respectively. Simply, then, there are good people and bad people, and this moral disparity is controlled by their respective possession of good and bad genes. Konrad Lorenz would, several years later, make a corresponding claim (see Chapter 3).

Should society form social policies based on the view that moral worth, or the lack of it, resides in people's hereditary endowment? Lenz apparently believed so. He drew what he saw as the third attribute of National Socialist racial hygiene and asserted: "The only way to eliminate genetic illness is through the negative selection of the afflicted families."[46] Thus, the genetic determinist idea that morality resides in the genes, and that environmental intervention cannot alter the evolutionarily selected presence of good genes in some people and bad genes in others, led in National Socialist racial hygiene ideology (as it did in the pre-Nazi writings of Haeckel and of Binding and Hoche) to the call for a negative eugenics "program," for genotype-destroying selections.

The appeal for the institution of selections to rid the world of diseased, immoral genes would become the hideous cornerstone of the Nazi program of applied biology.[47] For instance, as early in the Nazi regime as August 1933, Walter Schultze, the Bavarian commissioner of health, gave a speech in which he asserted that the sterilization of the racially inferior and the diseased was not a procedure secure enough to assume that the genes of these people would not insinuate themselves into the gene pool of the racially healthy. As a consequence, Schultze said that criminals, the mentally deficient, and other inferior people must be isolated and killed. To government officials of the Third Reich, it was clear very early that "the practice of extermination was part of the legitimate business of government."[48]

Although the concepts of the evolutionary importance of killing the unfit, of thereby replacing lost natural selections with man-made ones, and of thus halting the threat of domestication-induced degeneracy had a history in German racial hygiene thinking that can be traced from, at least, the arguments of Ernst Haeckel to those of Binding and Hoche, nineteenth-century writers other than Haeckel preceded Binding and Hoche by at least a quarter of a century. For instance, in 1895 Adolf Jost contended that the state's right to kill is the "key to the fitness of life," that to maintain the society and its health, the state must "own" death; it must adopt policies of killing to protect the racial health of the people.[49]

It is clear that, in the Nazi world view, disease and immorality were mutually defining. "Disease" for the Nazis was a quite broad concept, encompassing behavioral, moral, political, and cultural disorders or transgressions, as well as physical maladies.[50] Moreover, the "evidence" needed to diagnose a disease, particularly a genetic one, was quite broad. For instance, even non-Jewish, full-blooded German children were put to death (in what the Nazis termed a euthanasia program) because they had disorders ranging from severe mental retardation and physical deformities to such maladies as malformed ears, bed-wetting, or the mere label that they were difficult to educate.[51]

This conception of disease allowed Nazi ideologues to view biological goals (of racial survival and purification) and political goals (of destroying enemies of the state) as interchangeable. In an August 22, 1930, special issue of the *National Socialist Monthly* devoted to the celebration of the seventieth birthday of Alfred Ploetz, Theobald Lang wrote an article entitled

"National Socialism as the Expression of Our Biological Knowledge," in which he articulated what would later be echoed by Rudolf Hess, that National Socialism was merely applied biology. Lang presented what he construed to be the core principles of this version of biological "truth": that different people and races have distinct hereditary characteristics and that this genetic material was immutable, and that the current political and cultural context (the societal setting that existed in 1930, three years before Hitler assumed power) was functioning as a negative social selection.[52] As a consequence, he believed, the genetic quality of the German *Volk* was deteriorating.

The geneticist Eugen Fischer echoed this view. He contended that his science had demonstrated that "all human traits—normal or pathological, physical or mental—are shaped by heredity factors,"[53] and therefore he believed that human problem behaviors, and the threats to society they entailed, were genetic problems. With the collaboration of such scientists as Fischer, "The Nazis took problems of race, gender, crime, and poverty, and transformed them into biological problems. Nazis argued that Germany was teetering on the brink of racial collapse and that racial hygiene was needed to save Germany from 'racial suicide.' "[54]

Given this dire threat, the leader of the National Socialist state believed it was imperative to organize the German state to fight back against the enemies of the genetic health of the German people. To do so, the state must eliminate the possibility that the criminal, diseased genotypes will reproduce. Hitler, of course, accepted this duty. His 1935 signing into law of the acts known collectively as the Nuremberg laws was one relatively early attempt to use legislation to rid the German *Volk* of the threat of racial disease posed by these degenerate genotypes.[55] In addition, Hitler specified in some detail the dimensions of the biological challenge imposed by the presence of politically dangerous and genetically destructive genes within the German state:

> the *volkish* state must see to it that only the healthy beget children. . . . Here the state must act as the guardian of a millennial future. . . . It must put the most modern medical means in the service of this knowledge. It must declare unfit for propagation all who are in any way visibly sick or who have inherited a disease and can therefore pass it on. . . . Those who are physically and mentally unhealthy and unworthy must not perpetuate their suffering in the body of their children.[56]

With those words Hitler articulated his aim to rid the world of the "disease" of racial, moral, and political inferiority, and he would accomplish this end by erasing from the face of the earth the possibility that this disease could be transmitted. Because the disease was a genetic one, eliminating its transmission meant eliminating any possibility that the genes could be reproduced. The carriers of the genes must therefore be exterminated, and Hitler was not reluctant to announce this objective. For instance, using the possibility of war, which of course he was planning, as a pretext for this destruction of an entire group of people, Hitler announced on January 30, 1939, that, should war come, "the result will not be the bolshevization of the earth, and thus the victory of Jewry, but the annihilation of the Jewish race in Europe."[57]

Through this reasoning the policy of the destruction of a people was formulated. The path from prejudice to genocide was cleared for an entire nation.

THE SOCIAL APPLICATION OF GENETIC DETERMINISM IN NAZI GERMANY

In Hitler's Germany, as in the Germany of earlier decades, it was the Jew who was the carrier of genetic and political disease. It was Jews who were, within the *volkish* German state, the visibly sick, the possessors of inherited diseases, the bodily ill, and the spiritually unhealthy and unworthy. To the Nazis, Jews represented the epitome of the synthesis of genetic disease, moral depravity, and political criminality, and so it was Jews who threatened the purity of the *volkish* blood. And it was Jews and their blood about which Theodor Fritsch wrote his "Ten Commandments of Lawful Self-Defense":

> Thou shalt keep thy blood pure. Consider it a crime to soil the noble Aryan breed of thy people by mingling it with the Jewish breed. For thou must know that Jewish blood is everlasting, putting the Jewish stamp on body and soul unto the farthest generations. Thou shalt have no social intercourse with the Jew. Avoid all contact and community with the Jew and keep him away from thyself and thy family, especially thy daughters, lest they suffer injury of body and soul.[58]

The enormity of the problem presented by genetic disease in German society required more than the avoidance recommended by the commandments of Fritsch. According to Nazi ideology, the German state must, with singularity of purpose and unwavering commitment of all its "legal" and military powers, unite to fight against the onslaught of the subhuman carriers of this genetic disease. Although the mass murder that would be involved in waging such a battle would admittedly be a difficult task, all feelings of sympathy or pity for the Jews must be turned aside so that the greater glory of the "Thousand-Year Reich" would be maintained.

Emblematic of these sentiments were the remarks of Hans Frank, Nazi governor of the "General Government" (which covered major portions of Poland), in a speech on December 16, 1941: "But what should be done with the Jews? . . . We were told in Berlin '. . . liquidate them. . . .' Gentlemen, I must ask you to rid yourself of all feelings of pity. We must annihilate the Jews wherever we find them and wherever it is possible, in order to maintain here the integral structure of the Reich."[59] Similarly, Heinrich Himmler, addressing his senior SS (Schutzstaffel, or Defense Corps) officers at Poznan on October 4, 1943, indicated he would speak "quite frankly [on] a very grave matter . . . [which] should be mentioned quite frankly, and yet we will never speak of it publicly":

> . . . I mean . . . the extermination of the Jewish race . . . it's in our programme—elimination of the Jews and we're doing it, exterminating them. . . . This is a page of glory in our history which has never to be written and is never to be written. . . . We had the moral right, we had the duty to our people, to destroy this people which wanted to destroy us. . . . Because we have exterminated a germ, we do not want in the end to be infected by the germ and die of it. I will not see so much as a small area of sepsis appear here or gain a hold. Wherever it may form, we will cauterize it. Altogether, however, we can say that we have fulfilled this most difficult duty for the love of our people. And our spirit, our soul, our character has not suffered injury from it.[60]

Thus, the leaders of National Socialist Germany seemed quite certain that the only complete, or final, solution to the biological problem posed by the Jews could be one of genocide, to them a difficult but nevertheless noble task. Indeed, Hitler made sure that his murderous intentions were understood.

HITLER'S BIOPOLITICAL VIEWS:
GENES AS JUSTIFICATION FOR GENOCIDE

If National Socialism was only "applied biology," as Deputy Führer Hess proclaimed, then it was Hitler who provided the key theoretical linkage between the biological dangers threatening the German *Volk* and the political actions the Nazi state must take to counter this danger. On the one hand, Hitler was clear that there was a biological problem ("Anyone who wants to cure this era, which is invariably sick and rotten, must first of all summon up the courage to make clear the causes of this disease") and that he knew the source of the "infection."[61] The source of the sickness was the Jews. Following on the earlier characterizations of Jews by Paul de Lagarde,[62] Robert Lifton notes that in the Nazi era Jews were depicted as

> agents of "racial pollution" and "racial tuberculosis," . . . parasites and bacteria causing sickness, deterioration, and death in the host peoples they infested, . . . the "eternal bloodsucker," "vampire," "germ carrier," "people's parasite," and "maggot in the rotting corpse."[63]

But Hitler did not cast this biological threat only in biological terms. The threat was also one of political power: a degenerate biological group was trying to destroy a biologically superior one. Hitler was quite clear about his biopolitical view of Jews and of the political threat they posed to the German people. As one scholar summarizes Hilter's views, Jews were

> an antirace, formed out of a hybrid, indeterminate, mongrel core, a nomad people of eternal restlessness, incapable of independent political, territorial existence. . . . Their religion [is] a cover for their lust for unlimited power and for absolute rule over all others. Their control of the world [is] not based on territory, which they never had, and in this way they [differ] from all the other nations; it [is] based, rather, on financial and other machinations. . . . At first the Jew [demands] equal rights, and then, finally, superior rights, and . . . his aim [is] to rule the world; but as his character [is] parasitic and as he [is] incapable of separate existence, his rule would lead not only to the destruction of the nations oppressed by him, but also to his own demise.[64]

In short, for Hitler the political threat Jews posed was nothing short of the total destruction of any nation within which they resided. Such destruction would result also in the elimination of Jews, but if the nations housing Jews did not want to resign themselves to this dual annihilation, the only reasonable action would be to eliminate all Jews before they seized total control of the guest nation and took actions that would result in the death of the native populace. To Hitler, the issue was simply kill or be killed.

The killing of Jews as a self-defense against race annihilation was a concept adopted by German scientists adhering to Hitler's world view. For instance, Professor Eugen Fischer, co-author (with Erwin Baur and Fritz Lenz) of *Grundriss der menschlichen Erblichkeitslehre und Rassenhygiene* (The Principles of Human Heredity and Racial Hygiene) and director of the Kaiser Wilhelm Institute of Anthropology, Human Heredity, and Eugenics in Berlin-Dahlem, asserted in a June 20, 1939, lecture:

> When a people wants, somehow or other, to preserve its own nature, it must reject alien racial elements, and when these have already insinuated themselves, it must suppress them and eliminate them. The Jew is such an alien and, therefore, when he wants to insinuate himself, he must be warded off. This is self-defence. In saying this, I do not characterize every Jew as inferior, as Negroes are, and I do not underestimate the greatest enemy with whom we have to fight. But I reject Jewry with every means in my power, and without reserve, in order to preserve the hereditary endowment of my people.[65]

Thus, when the Nazis came to power they and the scientists supporting them could, as historian Yehuda Bauer puts it, extol murder "as a positive ethical command to save the world and white, Germanic supremacy because they accused the Jews of planning that fate for Germans" and therefore the Nazis "could murder because they accused the Jews of wanting to murder them."[66] Hitler believed it was the "cosmic" mission of the German people to destroy the inferior race challenging them and thereby to assume their predestined place at the top of the evolutionary hierarchy.[67] In other words, this mission for the German *Volk*, which Hitler would lead, had from at least the days of Haeckel been intertwined with a pseudoscientific evolutionary "story" of the genetic superiority of the *Volk* and the lesser value of, and threat imposed by, the lives of the other, subhuman races.

This conception culminated in the belief that the genocide of the Jews was a mission required for the salvation of the world. In short, the Nazi version of fascism, coupled with a biological determinist view of mystical and cosmic *volkish* racial superiority, led ultimately to the view that genocide was the only fully adequate means to save the German people from racial annihilation. As stated succinctly by Hitler's secretary, Martin Bormann, "National-Socialist doctrine is entirely anti-Jewish, which means anti-Communist and anti-Christian. Everything is linked within National Socialism and everything aims at the fight against Judaism."[68]

Genocide of the Jewish people played a crucial, enabling role in the Nazis' fulfillment of their "biosocial mission," a mission to apply the instruments of the National Socialist state in the service of the vision of racial superiority and of race war between the Aryans and the Jews. Given this centrality of genocide in their enactment of the biological determinist ideology, one can see what is behind Ernst Nolte's claim, in his *Three Faces of Fascism*, that "Auschwitz was as directly included in the principles of the racial doctrine of the Nazis as the fruit in its seed."[69]

Similarly, Raul Hilberg, in his *Destruction of the European Jews*, sees an inexorable progression toward genocide in the Nazis' translation of their racist ideology into social policies.[70] Four stages of policy were evident:

1. The *definition* of *volkish* racial superiority and of the biological threat of degeneration and annihilation posed by the genetically inferior Jews. As historian Saul Friedlander explains, this stage involves "Hitler's threats of extermination, which began at the end of 1938 and were expressed in well-known discussions with foreign dignitaries and in public speeches such as that of January 30, 1939, as well as in discussions with close aides after the defeat of Poland."[71]
2. The *expropriation* of Jewish prosperity (and of their civil and human rights). Such actions were coupled with policies expelling Jews from Germany and requiring them to emigrate; ideas to send Jews to some isolated location (for instance, Madagascar) were discussed in this context.[72]
3. The *concentration* of Jews (in the ghettos and camps in the "General Government" of Eastern Europe).
4. The *extermination* of Jews, beginning with various small-scale actions against Jews (and other groups) such as the so-called "euthanasia" program[73] and the Einsatzgruppen actions and ending with the genocide of the *Endlösung*.

Throughout this sequence there is evidence for the link between Hitler's racial, biological determinist ideology and Nazi policy. For example, in a public speech in Nuremberg in 1929, Hitler proclaimed that "the destruction of life unworthy of life" was a "worthy aim."[74] Indeed, within a few days after the beginning of World War II he issued a secret order for secret implementation of the "euthanasia" program.

But not all of Hitler's policy enactments were secret. The Nazis' enunciation of the biopolitical threat posed by Jews had been so thoroughly inculcated in the people of National Socialist Germany that a series of quite explicit anti-Jewish laws was readily enacted and obeyed.

THE LEGALIZATION OF ANTI-SEMITISM

When Hitler assumed power on January 30, 1933, he embarked on a series of actions designed to isolate the "Jewish disease" from the body of the *Volk* by legal means. Hitler had promised German President Paul von Hindenburg a legal solution to the "Jewish problem." On April 7, 1933, the first anti-Jewish law was enacted within Nazi Germany, and between that date and the end of the Third Reich, just over twelve years later, National Socialist Germany enacted more than four hundred laws and decrees aimed at the total elimination of Jews from Europe. The first act, the "Law for the Restoration of the Professional Civil Service," excluded Jews (and political opponents of the Third Reich) from the civil service. Another law, enacted at the same time, excluded "non-Aryans" from the legal profession (as either lawyers or judges) and from acting as jurors.[75]

Later in April 1933 the proportion of non-Aryans permitted in German schools or institutions of higher education was severely curtailed. By the end of October, non-Aryans and their spouses were excluded from government employment, barred from cultural and entertainment activities, and severely constrained in the contributions they could make to newspapers. The application of such laws was facilitated by an earlier decree, of April 11, 1933, which provided a legal definition of a non-Aryan: anyone "descended from non-Aryan, especially Jewish, parents or grandparents. . . . This is to be assumed especially if one parent or grandparent was of Jewish faith."[76]

Laws specifically directed against Jews soon followed. An April 21, 1933, law banned Jewish rituals involving the slaughter of animals for food. On July 14, Eastern European Jews who had become naturalized German citizens had that status canceled because they now were defined as political undesirables. And on September 25, Jews were denied the right to inherit farmland.[77]

The laws of 1933, however, were only a faint omen of the more severe laws that would follow. Beginning on September 15, 1935, on the occasion of the annual National Socialist party congress in Nuremberg, another set of anti-Jewish laws were enacted. The purpose of these so-called "Nuremberg laws" was to make the "purity of the German blood" a legal goal of the state.[78] Indeed, the "Blood Protection Law," passed on September 15, forbade marriage or even sexual relations between Jews and Germans and the "Marital Health Law" of October 18 required a certificate of health, racial and otherwise, before a marriage license could be obtained. Finally, the "Reich Citizenship Law" distinguished between "citizens" (that is, Aryan Germans) and "inhabitants" (that is, non-Aryans and, quite interestingly, unmarried women) and deprived Jews of all civil rights.[79]

No matter how comprehensive these laws isolating Jews and depriving them of civil rights may have seemed, Hitler did not believe they were enough to end the threat of the Jewish "disease." Historian Karl Schleunes explains:

> Simply to exclude Jews from certain professions, or to limit their numbers in the schools and universities, while it struck hard at the Jews, did little to further the biological separation demanded by a racially defined anti-Semitism. The entire Nazi ideological structure rested on the belief that the Jew was evil because of his blood. The promise of the Twenty-Five Points, of *Mein Kampf,* indeed of all of modern anti-Semitism, had been to end race-mixing.[80]

So still other action was required—because the threat the disease of the Jews posed went beyond even the dire political boundaries Hitler and his cohorts had specified. The "war against the Jews"[81] was not only a political war; it was a holy war as well.

GENOCIDAL POLICY AS A HOLY CRUSADE

Once Hitler had demonstrated the "courage" to identify the Jews as the cause of the problems in German society and as the threat to its future survival, the cure was obvious. Hitler said one "must call eternal wrath upon the foul enemy of mankind as the originator of our sufferings. . . . The fight against him becomes a gleaming symbol of brighter days, to show other nations the way to . . . salvation."[82] Under Hitler, this mission took the form of a religious crusade against the forces of Satan.[83] Thus, in *Mein Kampf* he wrote:

> In his vileness he [Satan] becomes so gigantic that no one need be surprised if among our people the personification of the devil as the symbol of all evil assumes the living shape of the Jew. . . . What we must fight for is to safeguard the existence and reproduction of our race and our people, the sustenance of our children and the purity of our blood, the freedom and independence of the fatherland, so that our people may mature for the fulfillment of the mission allotted it by the creator of the universe. . . . I believe that I am acting in accordance with the will of the Almighty Creator: by defending myself against the Jew, I am fighting for the work of the Lord.[84]

In short, Hitler's campaign to destroy the Jews was, to him, no less than a struggle of innate good versus innate evil, of God versus the devil. It is for these reasons that Hitler believed "no one need be surprised if among our people the personification of the devil as the symbol of all evil assumes the living shape of the Jew."[85]

Hitler's casting of the "Jewish problem" in religious terms was not unique within German society. This view can be traced at least to Martin Luther, whom Hitler cited as a great statesman who held views consistent with his own.[86] In this regard, Luther said: "Know, Christian, that next to the devil thou hast no enemy more cruel, more venomous and violent than a true Jew."[87] Indeed, to Luther, as later with Hitler, Jews *were* the devil or at least the "children of the devil."[88]

On biological, political, and religious/moral grounds it was therefore the duty of the Nazi state to exterminate all Jews, to cleanse the world of all traces of their being and of their religion. Policies, both covert and overt, had to lead to one ultimate aim: the total destruction of world Jewry. Only through the successful enactment of such policies could the biological vision of *volkish* racial superiority—so long enunciated in German history and epitomized in the pronouncements of Hitler—be fully realized.

THE ENACTMENT OF GENOCIDE

In accordance with the "Führer Principle,"[89] Hitler demanded total obedience to his will. But more than this principle operated to couple Nazi biopolitical and religious prejudice toward the Jew with genocidal political policies and direct, murderous *actions*. The Nazi system that Hitler and his cohorts devised was essentially one of direct application (of biology, as Hess claimed), of direct social action.

For instance, historian Shulamit Volkov suggests that whereas Hitler may not have had a definite plan for dealing with the Jews if and when he came to power, his anti-Semitism was always completely oriented to action.[90] In other words, Hitler's writings and speeches were not *just* political hyperbole, statements aimed *only* at bringing him political power. They were direct prescriptions for actions to be taken by those in political power:

> Nazism was a spoken culture. Its language was all speech, with no literary dimensions, no privacy, no individuality. It was the language of demagogy, of declamations and shouts, with flags flying in the wind and the swastika constantly before one's eyes. It was a culture in which verbal aggression was *not a substitute for action but a preparation for it.*[91]

Hitler's biological determinist vision—of *Kampf* and of the threat to *volkish* superiority posed by the genetically diseased Jews—may have been the reason that his words became a direct prescription for action. Historian Christopher Browning argues:

> Hitler's anti-Semitism . . . was a deeply held obsession. Ideologically, it was the keystone of his *Weltanschauung*. Without his understanding of politics in terms of a "Jewish-Bolshevik" conspiracy and his understanding of history in terms of a social Darwinist struggle of races (in which the Jews played the most diabolical role), the whole edifice would collapse. Finally, Hitler gave expression to this anti-Semitism in violent threats and fantasies of mass murder. Indeed, for a man whose social Darwinism implied the final resolution of any conflict in terms of the survival of one adversary through the "destruction" of the other, and whose anti-Semitism was understood in terms of race, mass murder of the Jews was a "logical" deduction.[92]

In essence, then, Nazi ideology and policy were action-conceptions, mergers of biological determinism, politics, religious zealotry, and murder, actions taken to protect the present and future German state against the degenerating genetic and political threats the Jews represented.

The historical record is replete with examples of this translation of Nazi ideology into murder on a mass scale. The action-orientation of Nazi ideology can be found in the accounts of those charged with the actual commission of the murders. For example, Otto Ohlendorf, a group leader of the SS, lieutenant general of police, and director of the Reich Security Main Office, was brought to trial in Nuremberg for being a major figure in the perpetration of countless atrocities. For instance, in just one year the Einsatzgruppen (Special Action Group) soldiers he commanded murdered about 90,000 Jews and other civilians in the southern Ukraine.[93]

During his defense Ohlendorf argued that all Jews, no matter what their age or sex, were an innate threat to the security of National Socialist Germany. This was the case, he believed, because Jews had instituted bolshevism in the Soviet Union and because this political orientation was, as the Führer claimed, merely a subterfuge for gaining dominance over other countries and for diminishing the racial purity of Aryans.[94] Specifically, Ohlendorf testified that all Jews were to be exterminated "not on account of their faith, or their religion, but because of their human makeup and character." When asked why it was necessary for the Einsatzgruppen to kill even Jewish infants and young children, Ohlendorf gave what to him was the obvious answer: "It is very simple to explain if one starts from the fact that [the Führer's] order not only tried to achieve security, but *permanent* security, lest the children grow up and inevitably, being the children of parents who had been killed, they would constitute a danger no smaller than that of the parents."[95]

In response to additional questioning, Ohlendorf described the details of a typical Einsatzgruppen action:

> The *Einsatz* unit would enter a village or town and order the prominent Jewish citizens to call together all Jews for the purpose of "resettlement." They were requested to hand over their valuables and shortly before execution to surrender their outer clothing. They were transported to the place of executions, usually an antitank ditch, in trucks—always only as many as could be executed immediately. In this way it was attempted to keep the span of time from the moment in which the victims knew what was about to happen to them until the time of their actual execution as short as possible.
>
> Then they were shot, kneeling or standing, by firing squads in a military manner and the corpses thrown into the ditch. I never permitted the shooting by individuals, but ordered that several of the

men should shoot at the same time in order to avoid direct personal responsibility. Other group leaders demanded that the victims lie down flat on the ground to be shot through the nape of the neck. I did not approve of these methods . . . because both for the victims and for those who carried out the executions, it was, psychologically, an immense burden to bear.[96]

Murder was justified, then, by Ohlendorf and others like him because the survival of the German *Volk* was being challenged by the presence of a genetically degenerate, morally inferior, and politically criminal disease entity, the Jew. Because the conception of racial, moral, and political health or disease was expressed biomedically (the Jew as a parasite, cancer, tuberculosis, bacillus, or simply as "disease incarnate"),[97] it was the duty of National Socialist Germans to protect themselves against the disease, to murder the vermin before they murdered the *Volk*. "Jews, lice, typhus!"[98] was a common Nazi slogan, and the carriers of such disease were targeted for total destruction by the Nazi state.

Given the Nazis' broad conception of the meaning of the term "disease," it was clear that just as Jews were genetically predetermined to manifest a host of diseases, they were at the same time innately destined to commit crimes.[99] Moreover, there was no remediation for the Jew. In complete adherence to the doctrine of biological determinism, the Nazis contended that no environmental intervention could alter the essential, diseased nature of the Jew. A passage in the 1937 *Handbook for the Hitler Youth* reads: "Environmental influences have never been known to bring about the formation of a new race. That is one more reason for our belief that a Jew remains a Jew, in Germany or in any other country. He can never change his race, even by centuries of residence among other people."[100]

The equation within National Socialist ideology of racial degeneracy, mental inadequacy, moral inferiority, political criminality, and environmental immutability legitimated the destruction of the Jew on biomedical, psychological, religious, moral, criminological, and anthropological grounds. "To be Jewish was to be both sick and criminal."[101]

Being thus categorized, Jews, as well as people in several other categories (communists, Polish intellectuals and clergy, Gypsies, homosexuals, the mentally retarded and emotionally disturbed, the severely handicapped, and other "useless eaters"), would be designated as "life not worth living," as "life unworthy of life."[102] Thus, in actualizing the visions of Haeckel and of Binding and Hoche, the Nazis set out to make "war against the Jews," to use all means available to German society to seek out and destroy totally every vestige of this superordinate enemy.[103] As we have seen, the mission involved an integration of biology and politics. For instance, Rudolf Höss, commandant of Auschwitz, recounting Heinrich Himmler's description of the purpose of the camp, said: "Jews are the eternal enemies of the German people and must be exterminated. All Jews within our grasp are to be destroyed without exception, now, during the war. If we do not succeed in destroying the biological substance of the Jews, the Jews will some day destroy the German people."[104]

The enactment of genocide was the job not only of the commandants of the death camps. In the Third Reich it was ultimately the job of the Nazi doctor to implement the "final solution" to the "Jewish problem," and physicians in the Third Reich saw their goal as nothing less than that of the head of the Nazis' SS, Himmler.

THE ROLE OF NAZI PHYSICIANS IN GENOCIDE

Of all the professional groups in National Socialist Germany, Hitler claimed that in order to act on his ideology one group was critical: physicians.[105] Robert Jay Lifton, Benno Müller-Hill, and Robert Proctor have all shown that Hitler was more than successful in engaging the cooperation of vast numbers of German physicians in his genocidal enterprise.[106] Disproportionate numbers of German physicians joined the ranks of the Nazi party, and the SS as well. In fact, the National Socialist physician saw himself as a "biological soldier."[107]

For instance, Rudolf Ramm, a physician at the University of Berlin, elaborated on the orientation introduced by Ploetz and maintained that a National Socialist physician was no longer just a provider of care to the ill but a "cultivator of the genes," a "physician to the *Volk*," a "genetics doctor" (*Erbarzt*), a "caretaker of the race," a "politician of population," and, as already noted, a "biological soldier."[108] Similarly, Hans Lohr, another physician, asserted that "the health of the *Volk* stands above the health of the individual" and that Nazi physicians have a "holy obligation to the state."[109] According to Lohr, this obligation involved finding diseased genetic material and destroying it,[110] soldierly action that would in Ramm's view prevent "bastardization through the propagation of unworthy and racially alien elements . . . [thereby] maintaining and increasing those of sound heredity [and attaining the national goal of] keeping our blood pure."[111]

Eduard Wirths, chief physician at the Auschwitz death camp, devoted himself to three interrelated features of Nazi ideology: "the claim of revitalizing the German race and *Volk*; the biomedical path to that revitalization via purification of genes and race; and the focus on the Jews as a threat to this renewal, to the immediate, and long-term 'health' of the Germanic race."[112] Armed with this ideological devotion, Wirths was a key implementer of the Nazis' final, biomedical solution to the "Jewish problem" threatening the health of the *Volk*. He sought to cure the German race by killing all those seen as endangering it, and his work at Auschwitz embodied the perverse Nazi inversion of healing through killing.[113]

The grotesque nature of the killing that took place at the death camps exceeded even that seen with the Einsatzgruppen. Testimony at the Nuremberg trials of a survivor of the Treblinka death camp, Samuel Rajzman, describes events that occurred when women, on arrival at the camp, were shaved completely (so that their hair could be used in Germany to, for instance, stuff mattresses):

> Because little children at their mothers' breasts were a great nuisance during the shaving procedure, later the system was modified and babies were taken from their mothers as soon as they got off the train. The children were taken to an enormous ditch; when a large number of them were gathered together they were killed by firearms and thrown into the fire. Here, too, no one bothered to see whether all the children were really dead. Sometimes one could hear infants wailing in the fire.
>
> When mothers succeeded in keeping their babies with them and this fact interfered with the shaving, a German guard took the baby by its legs and smashed it against the wall of the barracks until only a bloody mass remained in his hands. The unfortunate mother had to take this mass with her to the "bath." Only those who saw these things with their own eyes will believe with what delight the Germans performed these operations; how glad they were when they succeeded in killing a child with only three or four blows; with what satisfaction they pushed the baby's corpse into the mother's arms.
>
> The invalids, cripples and aged who could not move fast were put to death in the same way as the children. The ditch in which the children and infirm were slaughtered and burned was called in

> German the "Lazarett," "infirmary," and the workers employed in it wore armbands with the Red Cross sign.[114]

The perverse use of the Red Cross sign in the midst of such cruel murders underscores the Nazi corruption of the healing professions: they are those professions that must engage in mass killings in order to "cure" the Jewish disease that threatened the health of the German people and the purity of German blood.

In short, the role of the Nazi physician-turned-soldier was to engage in a biological search-and-destroy mission and in so doing serve the synthesized racial and political goals of the German state. In fact, an emblem of the full synthesis of the political and medical goals of the state is the fact that the National Socialist physician believed that Hitler "not only had the power of a commander in chief in a political sense, but was also the highest ranking physician."[115] In other words, Adolf Hitler was viewed by medical and nonmedical citizen alike as the chief "doctor of the German *Volk*."[116]

This synthesis of political and biomedical objectives, which led to the use of individual killings and mass murder techniques to "cure" the disease threatening the *Volk*, caused few problems in the minds of National Socialist physicians. Because they saw as their primary professional mission the healing of the race, of the mystical *Volk*, and *not* the treatment of individuals, there could be belief in "healing through killing," in using the murder of individuals to cure the disease afflicting the *Volk*.[117] This conversion of medical care into killing was epitomized in the views of Fritz Klein, an Auschwitz physician responsible for the selection of countless thousands for the *Sonderbehandlung* of gassing and cremation: "I am a doctor and I want to preserve life. And out of respect for human life, I would remove a gangrenous appendix from a diseased body. The Jew is the gangrenous appendix in the body of mankind."[118]

Ending the lives of carriers of racial disease was, in short, the highest duty of the Nazi physician turned "biological soldier." And because the life of a Jew was a life unworthy of life, a life not worth living, a life devoid of value, ending it would actually be a humanitarian act. Such a conclusion serves to synthesize the Nazis' biological and political conception of the Jew through a deadly dialectic. The political legitimation for the destruction of the Jews was achieved by advancing the biomedical argument that Jews were a genetic illness that could be cured only by elimination. In turn, the genetic determinist idea that the Jew introduced a degenerating and environmentally immutable genetic disease into the health of the German *Volk* was legitimated by specifying that the political "mission" of the German people is to form "a state which sees its highest task in the preservation and promotion of the most noble elements of our nationality, . . . the most valuable stocks of basic racial elements . . . [and] raising them to a dominant position."[119]

This mutual legitimation between science and politics was not restricted to a set of psychotic political zealots or professionally marginal biologists or physicians. Quite to the contrary, the merger of biological determinism and politics was the very essence of the German National Socialist state and as such permeated every sector and level of personal and professional life. There is ample documentation that numerous and quite preeminent scientists became deeply involved in the Nazis' societal "applications" of biological determinist ideology,[120] ranging from passive acquiescence to racial hygiene policies to an active shaping of ideology and policy. The history of collaborations of scientists and physicians with National Socialism has been presented most significantly in Lifton's *Nazi Doctors* (1986), Müller-Hill's *Murderous Science* (1988), and Proctor's *Racial Hygiene* (1988). However, the history of one physician-scientist's involvement with National Socialism—that of Konrad Lorenz—bears recounting here (although his role in Nazi racial science has been described by Lifton, Proctor, and others[121]).

A focus on Lorenz is important for three reasons. First, his ideas continue to be a voice for biological determinist thinking and thereby help legitimate the conceptually compatible claims of contemporary sociobiologists. Second, his views have not only become part of the mainstream of "normal science" but are even more seen by many as the cutting edge, as exemplary science. (Lorenz was awarded in 1973 the Nobel Prize for Medicine or Physiology.) Third, his work provides a counterpoint to a scientific alternative to genetic determinist thinking—that is, developmental contextualism. It is important, then, to look at the work of Konrad Lorenz—physician, biologist, and former Nazi party member—both during the era of National Socialism and up to his death in 1989.

3

KONRAD LORENZ'S BIOLOGICAL DETERMINISM: THEN AND NOW

The ultimate wisdom is always the understanding of instinct.
—*Hitler,* Mein Kampf

I am by inheritance obsessed with eugenics.
—*Konrad Lorenz, 1974*

The biological determinism of National Socialist Germany permeated all sectors of that society. The Nazis' Social Darwinism, eugenics, and racial hygiene movements were both scientific and political enterprises, engaging most notably the energies of governmental, business, and military leaders, on the one hand, and academic, biomedical, educational, and legal professionals, on the other. The breadth and depth of this participation in the maltreatment and ultimately murder of millions led historian Max Weinreich to state:

> It will not do to speak . . . of the "Nazi gangsters." This murder of a whole people was not perpetrated solely by a comparative small gang of the Elite Guard or by the Gestapo, whom we have come to consider as criminals. . . . The whole ruling class of Germany was committed to the execution of this crime. But the actual murderers and those who sent them out and applauded them had accomplices. German scholarship provided the ideas and techniques which led to and justified this unparalleled slaughter. . . . [The participants] were to a large extent people of long and high standing, university professors and academy members, some of them world famous, authors with familiar names.[1]

It is, then, a great historical irony that this widespread participation in the National Socialist agenda was associated with relatively few instances of criminal conviction and even fewer of severe sentencing. As historian Robert Proctor has noted, "Those convicted of crimes against humanity were only a handful of those who had participated in the construction of Nazi racial science and policy."[2] Indeed, it has been documented that many people who participated directly in the killing of the "racially unfit"— physicians, nurses, concentration camp guards—were either not brought to trial or, if they were, were given relatively light sentences, and that many of the scientists and physicians who were involved in the advancement of Nazi racial hygiene ideology and policies were not only *not* tried but even allowed either to continue in or to return unimpeded to academic posts and/or medical practices. Many of these former card-carrying and committed Nazi scholars and physicians even played important roles in postwar Germany and Europe, again becoming respected and influential, as they had been during the National Socialist era.[3]

What is rare, however, are cases of former Nazi party members receiving worldwide acclaim of the highest order for scientific theories that, at their core, remain indistinguishable from those promulgated during the Nazi era. One quite striking instance of this has occurred, however, in regard to the acclaim given the work of Konrad Lorenz, who became a Nobel laureate in 1973. The continuity between Lorenz's Nazi-era and post-Nazi theoretical positions is strong. The connections between the key theoretical ideas Lorenz continued to forward, through the time of the writing of this chapter, and the core conceptions in contemporary sociobiological thinking are strong as well. Indeed, historian Robert Richards contends that Lorenz "gave conceptual and empirical shape to the modern science of ethology, the science which has been further elaborated into (and . . . Wilson believes absorbed by) sociobiology."[4] To understand these linkages it is useful to discuss the continuing and career-long themes in the conceptual and empirical work of Konrad Lorenz.

CONTINUING THEMES IN THE WORK OF LORENZ

Lorenz was the foremost proponent of a branch of biology termed ethology, which involves the study of the evolutionary and thus hereditary bases of animal behavior. With two other eminent ethologists—Nikolaas Tinbergen and Karl von Frisch—Konrad Zacharias Lorenz was awarded the Nobel Prize for Medicine or Physiology in 1973 for his ethological theory and research regarding instinctual behavior in animals, particularly precocial birds, birds that walk or swim efficiently immediately after hatching. Lorenz used the term "imprinting" to describe such birds' social attachment to, and following along after, the first moving object they saw after hatching. Usually this "object" was a member of their own species, typically their mother, and the newly hatched birds would follow after this other bird; upon reaching sexual maturity later in their lives, the birds would try to mate with another such social object, in this case another bird of their own species.

Lorenz attempted to demonstrate the fixed nature of imprinting by showing that following and, later, mating behaviors could be directed to virtually any living organism, as long as it was the first moving object a newly hatched bird saw in the first critical hours after its birth. Birds could be made to imprint on boots or even on Lorenz himself. Indeed, numerous introductory psychology texts accompany their discussions of imprinting with a photograph of a somewhat stooped and kindly appearing Dr. Lorenz, an elderly man with white hair and beard, being followed by a troop of young ducklings.

According to Lorenz, the imprinting phenomenon was an instance of instinctual behavior in animals. Beginning in his earliest publications, Lorenz was concerned with the concept of instinct. Following the "discovery" by Oskar Heinroth, Lorenz noted "that there are motor patterns of constant form which are performed in exactly the same manner by every healthy individual of a species." He presented five criteria for determining whether an observed pattern of behavior reflected "inherited drives of fixed behaviors."[5]

Lorenz's criteria for an instinctual behavior pattern (as historian Robert Richards notes[6]) were (1) appearance of the behavior pattern in virtually all individuals of a species; (2) appearance of the behavior pattern in species members who were reared in experimentally controlled isolation, that is, who were (purportedly) deprived of experience; (3) complexity of the behavior pattern, that is, the learning capacity of the individual should not be sufficient for the acquisition of the behavior pattern and yet the behavior pattern was present; (4) appearance, or "release," of the behavior pattern either at inappropriate times or in incomplete ways (e.g., a bird may try to build a nest outside of the mating season); and (5) fixity and rigidity (that is, stereotypy and nonplasticity) of the behavior pattern—in other words, the behavior pattern took the same form whenever it appeared.

According to Lorenz, an instinctual behavior pattern could occur in one of two ways. First, an instinct could be observed when the individual experienced a specific "releasing" stimulus—that is, when the organism encountered a certain stimulus that "triggered" a given instinct—or, second, the instinct could be released in a seemingly spontaneous manner.[7] To explain these bases for the occurrence of an instinctual behavior pattern, Lorenz posited the existence of an "innate releasing mechanism" (IRM), a hypothetical mechanism believed to involve a set of receptor cells that released the instinctual behavior pattern when activated by a specific environmental stimulus.[8]

More specifically, Lorenz saw as instinctual certain inherited properties of nervous system structures. Some groups of neurons, he claimed, have specific instinctive properties built into them.[9] The structures obtain these properties directly from the organism's genetic inheritance, from the "interaction of the species with its environment during evolution . . . [that is,] by mutation and selection, a method analogous to learning by trial and success."[10]

To Lorenz, then, experience over the course of an organism's life (its ontogeny) has no role in the shaping, the development, of this neural structure. Instead, as comparative psychologist Daniel Lehrman points out, the key innate feature of such a neural structure is "its ability to select, from the range of available possible stimuli, the one which specifically elicits its activity, and thus the response seen by the observer."[11]

A classic example of a "fixed action," or instinctual, pattern deriving from an IRM involves the male three-spined stickleback fish. When this fish encounters another male three-spined stickleback with a red belly, the fish displays a set of behaviors indicative of threat; however, when the fish encounters a female with a swollen (but nonred) belly, the male displays the behavior pattern indicative of mating. Similarly, greylag geese will display a fixed action pattern involving escape responses when they encounter a white-tailed eagle, the only flying predator that is a danger to these geese.[11] However, the instinctual escape response can be released also if the goose is exposed to any object gliding slowly and silhouetted against the sky.[12]

Thus, more than one specific stimulus can engage the IRM and release an instinct. Indeed, as noted, fixed action patterns can occur "spontaneously"—that is, if the appropriate releasing stimulus has not been encountered for some period of time, then (apparently because of an accumulation of energy associated with the instinct and/or the IRM) the fixed action pattern "might go off in vacuo, as if dammed energy burst through containing valves."[13] Because of this "spontaneous release" feature in IRMs, Lorenz came to view instincts *not* as sets of reflexes but as drives, as constructs having motivational properties—that is, Lorenz believed animals sought out stimuli that would release their instincts, which would dissipate the energy associated with the instincts that had presumably been built up.[14] In other words, Lorenz was saying that, unlike a reflex, wherein one behaves automatically, instincts have motivational properties: instincts drive one to engage in particular behaviors.

Throughout his writings, Lorenz did not divorce himself from a commitment to the IRM concept, or from the belief that certain stimulus conditions can release, even in humans, a quite complex fixed action pattern—for example, involving aggression and "militant enthusiasm."[15] However, his concept of instinct evolved in ways other than changing from a reflex-like construct to a motivational construct. Lorenz changed his conception of "instinct" at least in part in response to criticisms from several comparative psychologists, who complained, for instance, about the artificial and simplistic distinction Lorenz drew between instinct and learning.[16] Lorenz often seemed to equate or subsume all nonevolutionary experience with the term "learning," when there are actually many more ways experience can influence behavior. For example, diseases, natural events (storms, earthquakes), wars, dietary regimens, and social and political policies, laws, and movements can all influence behavior even though those phenomena are not readily subsumed under the concept of "learning."

Scholars also objected to problems with experimental isolation studies.[17] For instance, in such studies the researcher can deprive the animal of only *some* experiences, since it is not possible to deprive a living organism of all experience; even a dark box is an environment, albeit a perhaps noxious one. Thus, an isolation experiment can tell us only that a particular experience is not necessary for a certain behavior; we can never determine from such a procedure that experience per se is not involved.[18]

Lorenz's concept of instinct was also criticized for ignoring the problem of development across life, at least as far as the presence of the IRM was concerned. The issue of how genes, which are chemicals, interact with cells, tissues, organs, and the environment to "build" across life the neural structures involved in the IRMs is never adequately discussed by Lorenz.[19] Furthermore, his concept of IRM was seen as problematic because it had an element of nonfalsifiability to it: an instinct was released either under specific environmental stimulus conditions *or* in the absence of them.

In response to such criticisms, Lorenz contended in his 1965 book, *Evolution and Modification of Behavior*: "What is preformed in the genome and inherited by the individual is not any 'character,' such as we can see and describe in a living organism, but a limited range of possible forms in which an identical genetic blueprint can find its expression in phenogeny. . . . The term 'innate' should never, on principle, be applied to organs or behavior patterns, even if their modifiability should be negligible."[20] Thus, Lorenz was now arguing that it is not the behavior pattern, no matter how fixed it may seem, which is instinctual; rather, it is the "range" of the behaviors involved in the instinct which is innate. In other words, what is preformed in, built into, the genome (the set of genes received, or inherited, at conception) is *information* about the forms of a given behavior that are possible for the species. Moreover, this conception of "instinct as information" requires the involvement of the environment in order for the presence of a behavior within the range of the instinct to be in evidence. This need for "nurture" in order to identify features of "nature" arises because, Lorenz contended, the environment is involved in the " 'decoding' of genome-bound information . . . [and as a consequence] contrasting of the 'innate' and the 'learned' as mutually exclusive concepts is undoubtedly a fallacy."[21]

Lorenz believed that by conceptualizing the information genes contain about such behavior patterns, instead of fixed action sequences, as innate he had addressed the criticisms of his purportedly simplistic division between "instincts" and "learning." Of course, it is not possible to observe this innate information. One can see only behavior, which, Lorenz now contends, is not instinctual per se but rather the product of "morphological ontogeny producing structure . . . and . . . of trial-and-error behavior exploiting structure as a teaching apparatus."[22] As a consequence, then, of his new (1965) conception of instinct, Lorenz replaced one set of problems with another. He gave up the five observational criteria for identification of instinctual behavior patterns and defined instinct as a nonempirical construct, which cannot be known directly. Information about the range of forms can only be inferred from behaviors that, by his own insistence, are not (however) independent of experience, or "learning."

The changes Lorenz introduced into his conception of "instinct" in 1965 led him away from the stance of an empirical scientist. His statements about the instinct concept became increasingly less linked to clear observational criteria and, as such, took on more of the character of an intuitive construct.[23] Nevertheless, an interest in the instinct concept spanned Lorenz's entire career, and he continued to be an advocate for the importance of the construct for understanding *both* animal and human behavior. Indeed, this advocacy was based on Lorenz's belief that human civilization was at risk because of phenomena associated with certain types of changes in instinctual patterns.

As a committed Darwinist, Lorenz believed that instincts were shaped by evolution, by natural selection. Instincts—whether defined behaviorally or through reference to genome information—afford, then, survival; their function is to allow the organism to "fit" the demands of the natural environment within which the particular set of behaviors, or range of forms, comprising the instinct has been selected. In this vein, Lorenz contends:

> Some information underlying an individual's behavior has indeed been "preformed" by the species. . . . It becomes all too easy to overlook the survival function of behavior altogether and, therewith, the selection pressure which caused its mechanisms to evolve. To anyone tolerably versed in biological thought, it is a matter of course that . . . any function of . . . survival value . . . must necessarily be performed by a very special mechanism built into the organic system in the course of its evolution.[24]

Therefore, according to Lorenz, instincts allow the organism to survive within the natural surroundings within which the instinct has been selected. The essence of instincts is to allow a fit with the demands of the environment within which the instinct evolved.

But what happens if and when individuals are taken from the environments within which their instincts have been naturally selected? What are the implications for survival when an organism finds itself in a setting other than the one within which its instincts evolved to fit? It is with these questions—involving the implications of taking individuals out of their natural selection environment and placing them in a tamer, more domesticated, more civilized setting—that much of Lorenz's scholarship, beginning in the 1930s and engaging him for much of the remainder of his life, was concerned.

That is the theme in the work of Lorenz on which the philosopher Theodora Kalikow focuses. Kalikow adopts this concern because of her view that Lorenz interweaves political ideology with his scientific focus, his discussion of the nature of instincts among domesticated animals or among humans "encountering" civilization. Kalikow states:

> Ideology played a triple role in Lorenz's speeches and writings during the years from 1938 to 1943. (1) He saw changes in the instinctive behavior patterns of domesticated animals as symptoms of decline. (2) He assumed a homology between domesticated animals and civilized human beings, that is, he assumed there must be similar causes for effects assumed to be similar, and he further believed that civilization was in a process of "decline and fall." Finally, (3) he connected the preceding concerns to racial policies and other features of the Nazi program.[25]

Kalikow also draws a distinction between Lorenz's early and later work:

> As examination of Lorenz's writings from before and after World War II shows that (1) and (2) have remained as features of his work, while (3) has disappeared, at least in its overt manifestations.[26]

In other words, then, Lorenz offers the hypothesis that human biological degeneration has been brought about through domestication. According to this hypothesis, the instinctual behaviors of "civilized," urban human beings, behaviors that evolved to fit more rural settings, have become increasingly more diseased and degenerate. Akin to the domestication-induced degeneracy that, he believed, afflicts animals reared away from their natural, or wild, setting, Lorenz contended that modern society's protection of humans from natural selection has resulted in the degeneration of human beings both intellectually and morally.[27]

Lorenz admitted, "I am by inheritance obsessed with eugenics."[28] Thus, if Kalikow's views are correct, the focus Lorenz adopted for his work on instincts is redolent of the Social Darwinist / racial hygienist thinking that was part of the intellectual, social, and political milieu of Germany before the Third Reich.[29] But are Kalikow's views of the conflation of science and politics in Lorenz's work correct?

To address this question let us first review the "evidence" Kalikow and others have marshaled in support of the contention that Lorenz's 1938–43 writings—his work during the Nazi era—combined science and politics.[30] Then we can look at and evaluate an analysis of Kalikow's views presented by Robert J. Richards, the historian who has been most critical of Kalikow's interpretation of Lorenz's Nazi-era writings. Finally, we will note the themes in Lorenz's writings that appear to have been carried beyond World War II and discuss the connection between the biological determinist views of Lorenz and those found among contemporary sociobiologists.

THE NAZI-ERA WORK OF LORENZ

Many of Lorenz's writings pertinent to his commitment to Nazi racist ideology and policies have been identified and translated by Theodora Kalikow, although other scholars have drawn attention to Lorenz's "Brown Past"—that is, his participation in Nazi party ("Brown Shirt") activities.[31] However, it was Kalikow who, in searching the records of the Berlin Document Center, found that Lorenz had applied for membership in the Nazi party on May 1, 1938, and was accepted (and given membership number 6170554) on June 28, 1938.[32] So Lorenz was both a scientist with doctoral degrees in medicine and zoology and literally a card-carrying member of the Nazi party. It seems reasonable to inquire, then, about the extent to which biological science and National Socialist ideology and policies were combined in Lorenz's work.

Throughout his career, Lorenz was concerned with the degeneration of instincts brought about by the domestication of animals with inferior genes. That theme has had time-honored status in German biological determinist / racist writings since at least the period of Haeckel and the Monist League. However, it was in 1938 that Lorenz first presented *his* views on how domestication was associated with human degeneration, at a meeting of the German Psychological Association in a paper entitled "Deficiency Phenomena in the Instinctive Behavior of Domestic Animals and Their Social-Psychological Meaning." First, Lorenz discussed the connection between instincts in animals and instincts in humans. Second, and sounding the theme first raised by Haeckel, Lorenz talked about how the domestication-induced degeneracy of instinctual behavior threatens the survival of the *Volk*. Third, he discussed how differences between the genetically fit and the genetically unfit (and degenerate) are manifested, one way being that fit people appraise beauty and aesthetic appeal as associated with the fit and not with the unfit. Fourth, and again similar to Haeckel, Lorenz argued that judgments of good and bad, or of moral or immoral, are associated with the hereditarily fit and the hereditarily unfit, respectively.[33]

For instance, with respect to the connection between instincts in animals and humans, Lorenz contended:

> What ought to be compared, in these inferences from animals to human beings, are the *hereditary* changes in the system of innate species-specific behavior patterns, changes that arise in animals in the course of domestication and in human beings in the course of the civilization process. These two processes, seen from the standpoint of the biologist, have much in common.

In regard to the threat to the survival of the *Volk* caused by this biological degeneracy, Lorenz stated:

> The similarity of the biological foundations makes it quite believable that these parallels, which extend to the smallest details of human and animal behavior, are not just superficial analogies, but are founded on underlying causes. Thus, through a closer investigation of the behavior of domestic animals, we may hope to further our understanding of the biological causes of many menacing decay phenomena in the behavior of civilized human beings.

Finally, in regard to the connection between innate goodness and badness and the presence of hereditarily fixed social behaviors that are either fit or not fit, respectively, Lorenz argued:

> Even the observer striving for complete objectivity cannot stop himself from evaluating the decay of social behavior patterns negatively, even in animals. This is even more the case with respect to our conspecifics. For humans we mean by "good" and "bad" really nothing other than "complete with respect to innate social behavior patterns" and the opposite of this. If a person in fact detachedly exhibits a thoroughly social behavior, but does this not according to feeling, or instinctively, but calculatingly, and we see through this, we never feel this person to be "good." Our instinctive evaluation thus really relates to the presence of absolutely specific *hereditary properties* in our conspecifics.[34]

Lorenz closed by cautioning that the carriers of "bad" hereditary properties can degenerate the health of the *Volk* "like the cells of a malignant tumor," a metaphor that Nazi racial hygienists used expressly for Jews, as the nonhuman antirace. Given this threat, Lorenz made what must be seen as the necessary call for social action, again quite similar to the appeals of other National Socialist physicians. Lorenz contended that those who possess fit instinctual patterns must capitalize on their aesthetic/valuational reactions to those who possess unfit, degenerating genes.[35] Using these reactions to recognize the unfit, the fit must eliminate the unfit in order to ensure the "racial health and power" of the *Volk*. Specifically, in closing his 1938 paper, Lorenz argued:

> This high valuation of our species-specific and innate social behavior patterns is of the greatest biological importance. In it as in nothing else lies directly the backbone of all racial health and power. Nothing is so important for the health of a whole Volk as the elimination of "invirent types": those which, in the most dangerous, virulent increase, like the cells of a malignant tumor, threaten to penetrate the body of a Volk. This justified high valuation, one of our most important hereditary treasures, must however not hinder us from recognizing and admitting its direct relation with Nature. It must above all not hinder us from descending to investigate our fellow creatures, which are easier and simpler to understand, in order to discover facts which strengthen the basis for the care of our holiest racial, volkish and human hereditary values.[36]

In short, as did Hitler, and Haeckel before him, Lorenz saw the mission of race purification—of protecting the *Volk* from the "malignant tumor" threatening it by the presence of genetic inferiors—to have cosmic, mystical, indeed holy characteristics.

In 1940 Lorenz published a paper entitled "Systematics and Evolutionary Theory in Teaching" in the journal *Der Biologe* (The Biologist), expanding on the themes of his 1938 paper and emphasizing, as Ploetz and other racial hygienists did, that the natural selection process had been eroded by modern civilization and that this erosion was the basis of the degeneration process threatening the survival of the *Volk*. He cautioned:

> Whether we share the fate of the dinosaurs or whether we raise ourselves to a higher level of development, scarcely imaginable by the current organization of our brains, is exclusively a question of biological survival power and the life-will of our Volk. Today espe-

cially the great difference depends very much on the question whether or not we can learn to combat the decay phenomena in Volk and in humanity which arise from the lack of natural selection. In this very contest for survival or extinction, we Germans are far ahead of all other culture-Volks.[37]

Given the nature of the journal in which the paper was published, it is not surprising that Lorenz sounded this caution about the need to combat the "decay phenomena" endangering the *Volk* and that he complimented his fellow Germans (although he himself was Austrian) for having the wherewithal to be winning this fight (through the racial policies of the National Socialist state). As Kalikow explains, *Der Biologe* was an organ of the Biology Section of the National Socialist Teachers' League, and its editorial board members came from such politically correct organizations as the National Socialist University Teachers' League, the SS, and the Race-Political Department of the Nazi party.[38] Given the striking similarity between the views expressed by Lorenz and those promulgated by Nazi physicians and politicians, and given the nature of the publication in which Lorenz presented his views, it is difficult to determine whether this 1940 article is a scientific statement or a Nazi political statement. If the article was meant as only a scientific statement, then at the very least one can wonder whether another publication outlet would have been more appropriate.

The article in *Der Biologe* is not the only one of Lorenz's 1938–43 publications to have a combined scientific and political message. In another paper published in 1940, "Domestication-Caused Disturbances in Species-Specific Behavior," appearing in the *Zeitschrift für angewandte Psychologie und Charakterkunde* (Journal of Applied Psychology and Personality), several reviewers of Lorenz's work, and Lorenz himself,[39] admitted that the most explicitly Nazi-oriented statements are made in regard to his interpretation of domestication-induced degeneracy. Indeed, Lorenz included at the end of that paper a section entitled "Practical Applications."[40] Throughout the paper, however, Lorenz repeated his themes of the danger of racial degeneration; the erosion of natural selection factors; and the need to applaud the National Socialist state's endeavors to institute their own selection measures and thereby exterminate the cancerous cells—the aesthetically ugly and the ethically evil—from the midst of the *Volk*. He stated that the paper addressed "the question whether the life-conditions of civilization and of domestic animal behavior contain factors that encourage mutations" and continued:

This problem receives its particular importance first through the knowledge that among the most dangerous and race-hygienically most damaging decay phenomena in the social behavior of civilized people are those which have their precise equivalents in the "domestication characteristics" of many domestic animals, and which, in all probability, depend on the same causes. On the answering of the question about these causes, however, depend the counter-measures to be taken. If there should be mutagenic factors, their recognition and elimination would be *the most important task of those who protect the race,* because the continuing possibility of the novel appearance of people with deficiencies in species-specific social behavior patterns constitutes a danger to Volk and race which is more serious than that of a mixture with foreign races. The latter is at least knowable as such and, after a one-time elimination of breeding, is no longer to be feared. If it should turn out, on the other hand, that under the conditions of domestication no increase in mutations takes place, but the mere removal of natural selection causes the increase in the number of existing mutants and the imbalance of the race, then race-care must consider an even more stringent elimination of the ethically less valuable than is done today, because it would, in this case, literally have to replace all selection factors that operate in the natural environment.[41]

Later in the paper, Lorenz expanded on this argument, especially in regard to how to deal with the threat posed by the "ethically less valuable" (a phrase strikingly redolent of Haeckel and, later, Binding and Hoche):

The only resistance which mankind of healthy stock can offer . . . against being penetrated by symptoms of degeneracy is based on the existence of certain innate schemata. . . . Our species-specific sensitivity to the beauty and ugliness of members of our species is intimately connected with the symptoms of degeneration, caused by domestication, which threaten our race. . . . Usually, a man of high value is disgusted with special intensity by slight symptoms of degeneracy in men of the other race. . . . In certain instances, however, we find not only a lack of this selectivity . . . but even a reversal to being attracted by symptoms of degeneracy. . . . Decadent art provides many examples of such a change of signs. . . . The immensely high reproduction rate in the moral imbecile has long been established. . . .

This phenomenon leads everywhere . . . to the fact that socially inferior human material is enabled . . . to penetrate and finally to annihilate the healthy nation. The selection for toughness, heroism, social utility . . . must be accomplished by some human institution if mankind, in default of selective factors, is not to be ruined by domestication-induced degeneracy. The racial idea as the basis of our state has already accomplished much in this respect. The most effective race-preserving measure is . . . the greatest support of the natural defenses. . . . We must—and should—rely on the healthy feelings of our Best and charge them with the selection which will determine the prosperity or the decay of our people[42] . . . [that is, charge them with] the extermination of elements of the population loaded with dregs. Otherwise, these deleterious mutations will permeate the body of the people like the cells of a cancer.[43]

Continuing the analogy between the presence of cancer cells within a body and the presence of a group of people within a society, Lorenz maintained:

There is a certain similarity between the measures which need to be taken when we draw a broad biological analogy between bodies and malignant tumors, on the one hand, and a nation and individuals within it who have become asocial because of their defective constitution, on the other hand. . . . Any attempt at reconstruction using elements which have lost their proper nature and characteristics is doomed to failure. Fortunately, the elimination of such elements is easier for the public health physician and less dangerous for the supra-individual organism, than such an operation by a surgeon would be for the individual organism.[44]

Because Lorenz called so clearly for reliance on the selection policies of "the Best" of Nazi Germany *and* for the extermination of elements of the population permeated with "dregs," *and* for a more severe elimination of the morally inferior, it is difficult to reconcile his claim of thirty-four years later: "That they meant murder when they said 'selection' was beyond the belief of anyone. I never believed the Nazi ideology. . . ."[45] It is puzzling, to say the least, that someone who called for "elimination" and "extermination" is surprised that those whose selection practices he congratulated exterminated those selected for eliminations. And for someone to claim that he never believed in Nazi ideology, when his publications make claims about biological determinism and call for social policies that dovetail precisely with the explicit details of such ideology, would seem to be a remarkable coincidence. The contradictions that might appear to exist between Lorenz's Nazi-era statements, and his postwar, later-life recollections of his wartime thoughts and meanings, are only compounded when we learn about his other publications during the Nazi period.

In a 1943 article entitled "The Innate Forms of Possible Experience," published in the *Zeitschrift für Tierpsychologie* (Journal of Animal Psychology), Lorenz reiterates his concept of the links between domestication-produced racial degeneration and aesthetic value judgments about what is ugly, and therefore what is threatening and dangerous (and hence "bad," in the moral sense) for society:

> If one systematically goes through the—on close observation—astonishing short list of the characteristics which clearly produce the ugly in human beings and animals, one comes to the result that they are all relational characteristics which in *human beings* indicate domestication- or civilization-caused decay phenomena. If the ugly is to be represented in art, the artist accordingly resorts, not to any old arbitrary distortions of the human ideal Gestalt, but with great regularity to the few typical characteristics of domestication. Classic Greek sculpture represented Silenus as the opposite of the god-and-hero-type, always pinch-headed, with pot-belly and too-short limbs, . . . and in just the same way the traditionally ugly Socrates is always pictured as a chondrodystrophic.[46]

The image Lorenz presents of the genetically degenerate and exemplary ugly person is remarkably akin to the depiction of the Jew presented in drawings found in Nazi-publisher Julius Streicher's rabidly anti-Semitic "newspaper," *Der Stürmer,* as well as in various elementary school primers and children's books Streicher published at the time Lorenz was writing these papers. Figure 1 presents two illustrations from Streicher's publications depicting Jews (in contrast to Aryans) as conforming closely to the characteristics Lorenz contended exemplified the domestication-induced degenerate ugly (for example, "pot-belly and too-short limbs"[47]).

Still other Nazi-era papers by Lorenz consistent with Nazi ideology and social policy appeared repeatedly between 1938 and 1943. For instance, another 1943 paper, "Psychology and Phylogeny," appearing in a volume edited by G. Heberer (*Die Evolution der Organismen* [The Evolution of

Figure 1. Taken from a racist primer published by Julius Streicher in 1936, *Trust No Fox in the Green Meadow and No Jew on His Oath,* these illustrations depict Jewish adults and children being expelled from a school (above) and from a town (below) as Aryan children look on and/or jeer. As is typical in Streicher's publications, Jews are drawn as potbellied and as having limbs that are too short, features that Lorenz claimed exemplified the domestication-induced degenerate ugly. Illustrations reproduced in Time-Life Books, *The New Order* (Alexandria, Va.: Time-Life, 1989), 105.

Organisms]), draws connections between domestication-induced degeneration phenomena in animals and humans and concludes that this "scientific"
evidence has clear and necessary racial-political implications:

> A domestic goose will mate nonchalantly with any gander, while the
> mating of the wild form is dependent on a vast quantity of compli-
> cated [and innate] "betrothal customs." In human beings, on the
> other hand, the expansion of [innate] schemata leads to the race-
> politically highly undesirable increase in the rate of reproduction of
> the inferior classes. . . . No inevitable "logic of time" brings the
> "senescence" of culture-nations with it, as Spengler believed—rather
> it is factors in the environment, which are concrete, accessible to
> experiment, and thus certainly possible to combat. The race-political
> necessity of their immediate, precise investigation is obvious.[48]

Such conflating of biological determinist racism and political necessities was
obvious both to Lorenz and to other contemporaries of his ideological
bent—"theorists" such as Hitler, German racial hygienists, and indeed any
good National Socialist who sought to justify through "scientific" argument
political actions aimed at the extermination of a people.

So from at least the time Lorenz joined the Nazi party in 1938 through
1943, when he entered military service for the Third Reich, Lorenz's papers
contained consistent themes that increasingly more clearly and stridently
appeared to combine his science with his racist-political views, views that
were entirely consonant with other statements by National Socialist scientists that merged politics and "scientific"-Nazi racial hygiene ideology.
Common among Lorenz's Nazi-era papers was the theme of domestication-
induced degeneracy; of aesthetically repulsive and immoral genetic misfits
multiplying at dangerous rates in society because of the erosion of natural
selection; of the need therefore to rely on "the Best" of the *Volk* to institute
selection measures to fight the threat to the race posed by these hereditarily
unfit "cancers"; and of the need for these state-designed selection measures
to involve elimination—extermination—of these degenerate "dregs."

The consistent repetition of these themes in several papers spanning a
half-decade cannot be interpreted as simply a temporary or minor aberra-
tion of a scientist toying with the implications of his work for political
ideology and social policy. It seems rather to be the work of a person
energetically explaining the important congruence between his science and
his politics, a person who wants to demonstrate to his audience how his

theory and research coalesce to give credibility to National Socialist biological determinist ideology and legitimacy to Nazi racial policies.

This conclusion is clearly predicated on the scholarship of Kalikow, who provided the seminal scholarship documenting the linkage between Lorenz's scientific views and National Socialist ideology.[49] However, Kalikow's view of this linkage has been questioned by Robert J. Richards in his authoritative and acclaimed book *Darwin and the Emergence of Evolutionary Theories of Mind and Behavior*.[50] It is therefore important to review and evaluate Richards's discussion of Kalikow's interpretation of Lorenz's Nazi-era work, in order to help clarify the nature of Lorenz's work during World War II and help us understand the ideas Lorenz presented after the end of the war.

EVALUATING THE "EVIDENCE" IN SUPPORT OF LORENZ'S NAZI PAST

According to Richards, Kalikow maintains that Lorenz's ideas about domestication-induced human degeneracy are tied both to the thinking of Haeckel and to the politics of the Nazis.[51] Influenced by the work of Gasman, who argued also that the Nazi "biological mission" was promoted by the Social Darwinism of Haeckel,[52] Kalikow finds in Haeckel (and other Monist League members), in National Socialist ideology, and in the writings of Lorenz four sets of ideas:[53]

1. A biological view of the world, a world in which the laws of nature and the laws of society are the same
2. The belief that human evolution has been moving with constancy until the present era, in which high reproduction rates and "humanistic" attitudes toward the less fit put the human race at risk for survival
3. The belief that there is a one-to-one relationship between outer human appearance and internal moral value (i.e., "what is beautiful is good") and that the Aryan race, which exemplifies the pinnacle of this correspondence, has its ancestry among the ancient Greeks
4. The idea that evolution is the creative force in the world, a notion that replaces the belief that God is the creator and shaper

Richards presents several reasons why the links Kalikow sees among the Haeckelian / Monist League views, National Socialist ideology, and Lorenz's ideas may be more apparent than real.[54] One reason is "nondistinctive-ness"—that the first two sets of ideas are present in the general literature on evolution whereas the last two sets were common at the turn of the nineteenth century.[55] Thus, Richards dismisses Kalikow's arguments regarding convergence among the three positions in regard to the four sets of ideas by contending, "If such vague similarities suffice here, we should all be hustled to the gallows."[56] One may question, however, whether the similarities Kalikow finds are as vague as Richards portrays. Furthermore, one may ask whether there is support for the convergence among the three positions in addition to the four sets of ideas noted by Kalikow. An examination of the other reasons that Richards rejects Kalikow's ideas will allow us to address these questions.

Richards's second reason for disagreeing with Kalikow's linkage of the three positions is that the intellectual influence of Haeckelian / Monist League views on National Socialist ideology was not a completely clear one.[57] That may certainly be the case, but there is little reason to expect that the hodgepodge of concepts, the opportunistic twisting of the motley set of ideas, that constitutes the corpus of Nazi ideology should show a neat and logical pattern of influence. For instance, Ralph Manheim, translator of the most frequently cited English version of *Mein Kampf,* indicates in his notes to that edition that Hitler never attempted to systematize his knowledge, that he relied largely on disjointed facts.

> Even where he is discussing theoretical matters like "the state," "race," etc., he seldom pursues any logic inherent in the subject matter. He makes the most extraordinary allegations without so much as an attempt to prove them. Often there is no visible connection between one paragraph and the next.[58]

In short, if Nazi ideology was not a logical and coherent system, it is not appropriate to make the presence of a coherent pattern of influence a criterion for linkage between Haeckelian / Monist League views and Nazi ideology.

Accordingly, in order to understand the "intellectual influences" on Nazi ideology, Hitler's presentation in *Mein Kampf* and the succeeding tracts by Nazi ideologues may best be scrutinized for the sources that (not necessarily logically or correctly) are reflected in them. For instance, whereas both Richards and Kalikow note that evolutionary theory was not fully accepted in the Third Reich,[59] it is known that Hitler was influenced by German

Social Darwinist / racial hygiene thinking. The eminent molecular geneticist Benno Müller-Hill notes that while Hitler was imprisoned in Landsberg in 1923 he read the second edition of the textbook by Baur, Fischer, and Lenz, *Grundriss der menschlichen Erblichkeitslehre und Rassenhygiene* (The Principles of Human Heredity and Racial Hygiene)[60] and subsequently incorporated racial ideas into *Mein Kampf,* which he was preparing during his imprisonment.

Thus, whereas all features of evolutionary thinking are not necessarily present in Hitler's writings, we do find the idea of selection by a wise but ruthless nature; the notion that a hardened race of high accomplishment will eventually emerge under such conditions; and the view, present in German Social Darwinist / racial hygiene writings, that societal interference with this process will permit the weak and the sick ("lives unworthy of life") to survive and the quality of the race to be thereby diminished.[61] Hitler wrote:

> Nature herself in times of great poverty or bad climatic conditions, as well as poor harvest, intervenes to restrict the increase of population of certain countries or races; this, to be sure, by a method as wise as it is ruthless. She diminishes, not the power of procreation as such, but the conservation of the procreated, by exposing them to hard trials and deprivations with the result that all those who are less strong and less healthy are forced back into the womb of the eternal unknown. Those whom she permits to survive the inclemency of existence are a thousandfold tested, hardened, and well adapted to procreate in turn, in order that that process of thoroughgoing selection may begin again from the beginning. By thus brutally proceeding against the individual and immediately calling him back to herself as soon as he shows himself unequal to the storm of life, she keeps the race and species strong, in fact, raises them to the highest accomplishments.
>
> At the same time the diminution of number strengthens the individual and thus in the last analysis fortifies the species.
>
> It is different, however, when man undertakes the limitation of his number. He is not carved of the same wood, he is "humane." He knows better than the cruel queen of wisdom. He limits not the conservation of the individual, but procreation itself. This seems to him, who always sees himself and never the race, more human and more justified than the opposite way. Unfortunately, however, the consequences are the reverse:

While Nature, by making procreation free, yet submitting survival to a hard trial, chooses from an excess number of individuals the best as worthy of living, thus preserving them alone and in them conserving their species, man limits procreation, but is hysterically concerned that once a being is born it should be preserved at any price. This correction of the divine will seems to him as wise as it is humane, and he takes delight in having once again gotten the best of Nature and even having proved her inadequacy. The number, to be sure, has really been limited, but at the same time the value of the individual has diminished; this, however, is something the dear little ape of the Almighty does not want to see or hear about.

For as soon as procreation as such is limited and the number of births diminished, the natural struggle for existence which leaves only the strongest and healthiest alive is obviously replaced by the obvious desire to "save" even the weakest and most sickly at any price, and this plants the seed of a future generation which must inevitably grow more and more deplorable the longer this mockery of Nature and her will continues.[62]

The link between the German Social Darwinists / racial hygienists and Hitler's ideology is underscored by the consistency between the recommendations Binding and Hoche made regarding the treatment of the weak, lame, and ill in their 1920 book *Die Freigabe der Vernichtung lebensunwerten Lebens* (The Sanctioning of the Destruction of Lives Unworthy to Be Lived) and Hitler's views in *Mein Kampf*. Hitler said:

It is a half-measure to let incurably sick people steadily contaminate the remaining healthy ones. This is in keeping with the humanitarianism which, to avoid hurting one individual, lets a hundred others perish. The demand that defective people be prevented from propagating equally defective offspring is a demand of the clearest reason and if systematically executed represents the most humane act of mankind. It will spare millions of unfortunates undeserved sufferings, and consequently will lead to a rising improvement of health as a whole. . . . The right of personal freedom recedes before the duty to preserve the race.[63]

Hitler turned those ideas into policy (for example, involving the "forced euthanasia" of German children) when he assumed power.[64] But to him—

and to at least some proponents of the Haeckelian / Monist League views as well as to Lorenz during both the Nazi era and two decades after it—forced-euthanasia programs are not the best way to "prune the weak" and move the race in the direction a wise and ruthless Nature would select. Rather, it is through warlike behavior—aggression, struggle, and killing of other humans—that such selection will occur.[65] Thus, according to Hitler:

> There will be but two possibilities[,] either the world will be governed according to the ideas of our modern democracy, and then the weight of any decision will result in favor of the numerically stronger races, or the world will be dominated in accordance with the laws of the natural order of force, and then it is the peoples of brutal will who will conquer, and consequently once again not the nation of self-restriction.
>
> No one can doubt that this world will some day be exposed to the severest struggles for the existence of mankind. In the end, only the urge for self-preservation can conquer. Beneath it so-called humanity, the expression of a mixture of stupidity, cowardice, and know-it-all conceit, will melt like snow in the March sun. Mankind has grown great in eternal struggle, and only in eternal peace does it perish.[66]

Similarly, Heinrich Ziegler, a founding member of the Monist League, argued:

> According to Darwin's doctrine war has been of the greatest importance for the general progress of the human race, since the physically weaker, the less intelligent, and the morally degenerate must make way for the stronger and better developed people. . . . If one accepts the insights of modern science, he must see war between different races or people as a form of the struggle for existence in the human race.[67]

Lorenz, in turn, noted:

> It is quite typical of man that his most noble and admirable qualities are brought to the fore in situations involving the killing of other men, just as noble as they are. . . . Aggression, far from being the diabolical, destructive principle that classical psychoanalysis makes

it out to be, is really an essential part of the life-preserving organization of instincts.[68]

Hitler's idea of racial greatness through eternal struggle, and Ziegler's
notion of racial war as a feature of the human race's struggle for existence,
converge with Lorenz's 1966 view that aggression is life-preserving and
that, by acting on his aggressive instincts, man [sic] has often attained
nobility and other admirable characteristics.

Such linkages among the ideas of Hitler, the Monist League literature,
and Lorenz are not consistent with Richards's view that only "vague
similarities" exist across the three positions.[69] Moreover, although Richards
points out that the Monist League had a pacifist, socially liberal orientation,
such a general stance does not gainsay either Ziegler's conception of the
race-preserving function of war or the possibility that individual scientists
may be personally committed to a pacifist political ideology and yet committed as scientists to a belief about the inevitable, or even instinctual, basis
of human aggression.[70] Indeed, this is just the stance Lorenz took in his
book *On Aggression,* in which he argues that to avoid the release of
instinctual "militant enthusiasm" society must find means to discharge
aggression in innocuous ways.[71]

In short, the linkages Kalikow draws appear to be real, perhaps even
beyond the extent posited by Kalikow.[72] For instance, although Kalikow
points out that the Nazis did not share with either Haeckel or Lorenz the
commitment to evolution per se as the creative force in the world, we have
seen in Hitler's *Mein Kampf* an emphasis in National Socialist ideology on
the selective and shaping force of nature *and* on several ideas associated
with German Social Darwinist / racial hygiene thinking, and quite notably
the notion of domestication-induced degeneracy. Indeed, Richards indicates
that some National Socialist ideologues did in fact eulogize Haeckel, credit-
ing him with providing scientific support for ideas central to the Nazis'
biologized view of the world.[73]

Nevertheless, although Richards himself provides some evidence for the
links between Nazi ideology and Haeckelian / Monist League views, he
offers two additional reasons why Kalikow is mistaken in making this dyad
a triad by adding the views of Lorenz. Richards points out that Lorenz never
cited Haeckel's work as supportive of his own and that Lorenz held that the
key facet of Haeckel's theory of heredity—the idea of the inheritance of
acquired characteristics—was scientifically unsound.[74] Neither of these objections to Kalikow's argument seems strong, however.

Kalikow's point seems to be more that Lorenz's ideas were consistent with the views found within the general orientation of Haeckel and the Monist League, and *not* that Lorenz adopted either all of these ideas or any of the ideas of a particular member of the league—including (as it seems) Haeckel. Indeed, the above presentation of converging quotes from Hitler, Ziegler, and Lorenz suggests that there is some correspondence between the views of Lorenz and at least one prominent member of the Monist League. In addition, both Richards and Kalikow point out that Lorenz's original formulation of "instinct" followed the conceptual lead of Ziegler.[75] Furthermore, Kalikow and, earlier, Nisbett have noted that Lorenz's early interest in evolutionary biology was prompted by his reading a book by Wilhelm Boelsche, co-founder of the Monist League, *Die Schöpfungstage* (The Days of Creation).[76]

Perhaps because of the nature of these last two points made in argument against Kalikow's position, Richards concludes his analysis of her position by interpreting the historical record of Lorenz's Nazi party affiliation and of his publication record (from 1938 to 1943) during the era of the Third Reich as "a gossamer thread by which to tie Lorenzian biology to the Nazis."[77] However, Richards does not deny that Lorenz wrote papers consistent with National Socialist ideology, but only that they represent his "few occasions of public Nazi association."[78] One must therefore decide how often a person must make a public commitment, in speeches and/or in writing, to a given ideology before that person can be justifiably linked with it. Perhaps what is operating in Richards's stance on this issue is a commitment to the Frankonian proverb "Amol schad' kan Malda nix," "Once does not do a maiden any damage."[79]

In any case, Richards does conclude that "Lorenz [in 1940] undoubtedly descended to accommodate some of his biological views to the ideology of his time and place" and that "at this point in Lorenz's career, certain well-entrenched evolutionary ideas happened to intersect with despicable Nazi dogma."[80] Thus, Richards appears to come full circle to admit Kalikow's point about the convergence of Lorenzian and Nazi "biology," and he leaves one to wonder only about the pervasiveness of the association and Lorenz's enthusiasm for it.

In regard to pervasiveness, we have the historical record provided by Kalikow, as well as by some other authors—for instance, Müller-Hill.[81] Regarding enthusiasm, it may be (as Richards speculates) that, had the Weimar Republic survived, the main features of Lorenz's work would have remained the same.[82] Indeed, we have seen that the racial ideas found in

Nazi ideology had a long history antedating the Third Reich, and that history might have involved Lorenz in some other manner had the events of 1933–45 not taken place. Nevertheless, it is difficult to gainsay Lorenz's enthusiasm for Nazi ideology when we learn that in his 1940 paper in *Der Biologe* Lorenz called it "one of the greatest joys of [his] life" to have converted a student to "our concept of the world (*Weltanschauung*)"—that is, to National Socialism.[83]

So there seems to be both appearance and reality to Kalikow's views of the linkage between Lorenzian writings during the Nazi era and the biologized world view of National Socialist politics.[84] Moreover, Kalikow's interpretation of the connection between themes in the writings of Lorenz and in Nazi ideology is underscored when the continuity between the key theoretical ideas found in Lorenz's Nazi-era writings and his post-Nazi-era work is recognized. It is therefore useful to focus on this continuity, and also to explain its important role in legitimating more contemporary biological determinist claims, such as those found in sociobiology.

LORENZ AFTER WORLD WAR II

Throughout his scholarly career, Konrad Lorenz maintained a central interest in the role of evolution, and of heredity, in animal and human behavior, focusing on the importance of instincts in understanding such behaviors as social attachment, aggression, and moral or ethical functioning. Morality may involve not only behaving in particular ways but also knowing right from wrong, good from bad. Given this connection between morality and knowledge, we can also understand Lorenz's career-long interest in the evolutionary basis of humans' knowledge, or of their cognitive system.[85]

All these themes in Lorenz's scholarly work are indisputably appropriate arenas for academic intellectual endeavor. His continuity of interest in these topics could be evidence of an admirable scholarly commitment to long-term programmatic research—research that would be regarded as sound *if* we were to judge it in isolation from the theoretical ideas from which it sprang. But if his work involved a merger of scientific and racist political purposes, his continuity of interest might tell us as much about enduring political agendas as about scientific ones. If little has changed in Lorenz's core scientific/political message, other than the deletion after World War II of Nazi terminology to present the message, then one must be skeptical (at the least) about Lorenz's 1974 claim that his involvement with the Nazis and with their ideology was merely a foolish, "naive error": "Like a fool I thought I could improve them [the Nazis and their ideology], lead them to something better. It was a naive error."[86] The issue we need to address, then, is (1) whether the message in the writing of the Nazi-era Lorenz was a combined scientific/racist one while the message after the Nazi era has been solely scientific, or (2) whether the messages have remained essentially the same, with only the Nazi-era terminology omitted.

If a case can be made for the second alternative, the implications for present-day science and social policy would be great, in light of the scientific eminence Lorenz enjoyed and the credibility and respect thereby accorded his work. That is, politicians and policymakers could get the impression that there is broad-based scientific acceptance of biological determinist thinking that at its core has pejorative, racist political aims. The resulting danger would be that these policymakers might promote social policies that are consistent with those particular assertions of biological determinism. For example, interpretations that the bases of criminality and immorality are genetically based would come to the fore, and devoting resources to environmental remediation programs would therefore not be seen as economically prudent.

The most straightforward way to decide about the continuity or discontinuity in Lorenz's core message is to refer to his own post-Nazi-era statements. Has the tune really changed, or has the song only been given a new name? Continuity alone in the topics Lorenz studies is not enough to make a decision; we must see whether his stance on the key theoretical theme in his work, if one exists, has been altered appreciably.

There is a core theme uniting Lorenz's Nazi-era and post-Nazi-era work: the threat posed by domestication-induced degeneration of human instincts for the survival and further evolution of human moral, or ethical, being and thus for the future survival or progress of civilization. It is my belief that Lorenz's current interpretation of this theme continued to be identical to that in his Nazi-era papers, reviewed earlier in the chapter. Lorenz may have claimed that he never believed Nazi ideology, that he sees his use of it as a "naive error," but he never apologized or claimed regret for the *interpretations* he made—which were consistent not only with Nazi ideology but also, it is important to add, with the ideas of pre-Nazi and Nazi-era German Social Darwinists, eugenicists, and racial hygienists, such as Haeckel, Ploetz, Schallmayer, Binding, Hoche, and Lenz. Indeed, the only specific facet of his message for which I can find an apology is his choice of *terminology*; he does not apologize for the underlying ideas the particular terms conveyed. Lorenz said: "In retrospect, I deeply regret having employed the terminology of the time . . . which was subsequently used as a tool for the setting of horrible objectives."[87] However, and perhaps revealing of the actual continuity in the core, underlying theme of his work, Lorenz indicated in 1974 that he was intrigued with eugenics to the point of obsession.[88]

We might expect to find, therefore, that the only changes in Lorenz's views about the threats posed to civilization by domestication-induced genetic degeneracy are in the way the views are phrased. While this is partly the case, one can also find a continuing ideological *and* terminological emphasis on changing the distortions of natural selection that modern civilization's domestication practices have wrought; on the genetic basis of morality, of the human sense of good and bad; on the decay brought about in this instinctual capacity by domestication phenomena; on the fact that some people have genes for good morality and/or ethics and that others have ethically bad genes; and on the inevitable need for elimination (if not extermination) procedures to protect society against further degeneration.

For instance, in an article appearing in 1954 Lorenz equated domestication phenomena with deleterious mutations and pathologies and indicated that civilization's interference with (or removal of) natural selections processes was responsible for the appearance of such phenomena:

> One might possibly be inclined to think that environmental conditions . . . have favored homologous mutations. However, this would definitely seem to be a false assumption; instead the blame for the appearance of these characters seems to be exclusively due to the removal of natural selection. . . . Domestication-induced alterations of instinctive behavior are, by nature, processes bordering closely on pathological events.[89]

Moreover, as in the Nazi era, the postwar Lorenz indicates that this domestication-induced genetic degeneracy occurs in humans as a consequence of modern civilization's interference with naturally selected instinctual behavior patterns. In 1950 he contended:

> With every organism that is plucked out of its natural environment and placed in novel surroundings, behaviour patterns occur which are neutral or even detrimental for the survival of the species. . . . Modern man represents such an animal, torn from his natural

environmental niche. . . . The flowering of human culture has so extensively changed the entire ecology and sociology of our species that a whole range of previously adaptive endogenous behaviour patterns have become not only non-functional but extremely disruptive.[90]

To Lorenz, then, instinctive behavior patterns arose as naturally selected adaptations to humans' premodern context. These instincts are fixed patterns of action; they are not flexible (or plastic) in and of themselves, and they are not available for modification either in or through the action of an altered environment. Thus, when humans find themselves in the radically new setting of modern civilization they are in twofold peril: (1) their previously adaptive instincts may no longer be useful in the new setting and (2) the removal of natural selection from the new setting will allow degenerative instincts (what Lorenz terms "deleterious mutations" and "pathologies") to survive and be reproduced.

THE EXAMPLE OF HUMAN AGGRESSION

Lorenz's views regarding aggression provide an instructive example of how an instinctual pattern that purportedly evolved to facilitate human survival may undermine it in the context of modern civilization. In his 1966 book, *On Aggression*, Lorenz describes humans' aggression as involving instinctual "militant enthusiasm," an inherited vestige of their past and an instinctual response that allowed the individual to respond, with confederates in his or her group, to threats from organisms outside the community.[91] Indeed, Lorenz sees such an instinctual pattern as one that, with no thought involved, allowed communities of even fully evolved humans to survive:

> To the humble seeker of biological truth there cannot be the slightest doubt that human militant enthusiasm evolved out of a communal defense response of our prehuman ancestors. The unthinking single-mindedness of the response must have been of high survival value even in a tribe of fully evolved human beings. It was necessary for the individual male to forget all his other allegiances in order to be able to dedicate himself, body and soul, to the cause of the communal battle.[92]

Lorenz contended that with changes in cultural development the "object" that is defended by the militant enthusiasm instinct may change as well. For example, in early human evolution the immediate group may have been the object toward which a threat would have elicited militant enthusiasm, whereas among contemporary humans the nation or an abstract idea (for instance, "democracy") may elicit the instinct.[93] Whatever the object, Lorenz believed two points were certain. First, the object that is salient in a culture becomes so "by a process of true Pavlovian conditioning plus a certain amount of irreversible imprinting," and second, culture owes a great debt to militant enthusiasm, because "without the concentrated dedication of militant enthusiasm neither art, nor science, nor indeed any of the great endeavors of humanity would ever have come into being."[94]

But although all these positive outcomes of civilization derive from the instinctual aggression of human beings, civilized humans are not, in Lorenz's view, entirely in control of whether these outcomes will materialize. Because militant enthusiasm is an instinct that attaches to an object through irreversible imprinting and reflex-like learning or conditioning in early life, negative outcomes of the instinct's release and attachment to a cultural object, outcomes such as war, may occur. Thus, in speaking of whether militant enthusiasm will in fact lead to positive social outcomes, Lorenz contended:

> Whether enthusiasm is made to serve these endeavors, or whether man's most powerfully motivating instinct makes him go to war in some abjectly silly cause, depends almost entirely on the conditioning and/or imprinting he has undergone during certain susceptible periods of his life. There is reasonable hope that our moral responsibility may gain control over the primeval drive, but our only hope of its ever doing so rests on the humble recognition of the fact that militant enthusiasm is an instinctive response with a phylogenetically determined releasing mechanism and that the only point at which intelligent and responsible supervision can get control is in the conditioning of the response to an object which proves to be a genuine value under the scrutiny of the categorical question.[95]

Lorenz, then, offers hope that, if civilization recognizes the instinctive nature of aggression, future generations can be attached to cultural objects subserving the most prized and positive achievements, and also the moral responsibility, of human beings. Indeed, in *On Aggression*, he describes what he believes are "simple and effective" ways of "discharging aggression" in an "innocuous manner" through attempting to "redirect it at a substitute object" and suggests that sports may be particularly useful in such attempts to channel militant enthusiasm in nondestructive ways.[96]

What if instinctual aggression is *not* redirected by civilization? What if militant enthusiasm is released and attached to an object that is associated with war? Given the instinctual, reflexive, and irreversible character that Lorenz attributes to human aggression, there is little a person or group can do if early experience leads militant enthusiasm to be associated with negative—dangerous and destructive—outcomes.

It is possible to view this "double-edged sword" character of human aggression in a historical context. On the one hand, the instinctual and reflexive character of militant enthusiasm can be controlled *in the future,* if society presents appropriate imprinting, conditioning, and redirection, so there is hope that aggressive instincts can subserve moral and positive aims. On the other hand, Lorenz's formulation excuses the past: if society did not recognize the evolutionarily determined, instinctual nature of aggression, and if therefore a cohort of people were exposed in their early youth to an inappropriate object, they are not morally culpable for having had their instinct released by this object. Knowing *now* that human aggression is instinctual may make the leaders of society morally responsible for building programs for the future that will involve the nondestructive release of instinctual aggression (e.g., through sports programs). However, current groups of adults cannot be blamed if leaders of the society they experienced as children did not act in this responsible manner. In short, it is possible to interpret Lorenz's formulation of instinctual militant enthusiasm as excusing the past—perhaps, more specifically, his past—while providing hope for the future.

This interpretation is bolstered when one reviews Lorenz's ideas about "the stimulus situation which releases" militant enthusiasm.[97] Lorenz contended that there were four stimulus conditions that led to the appearance of militant enthusiasm; and when militant enthusiasm appeared in this way, Lorenz believed it occurred with a degree of certainty equivalent to an inborn reflex, such as an eyeblink.[98] He argued:

> Militant enthusiasm can be elicited with the predictability of a reflex when the following environmental situations arise. First of all, a social unit with which the subject identifies himself must appear to be threatened by some danger from outside. . . . A second key

stimulus which contributes enormously to the releasing of intense militant enthusiasm is the presence of a hated enemy from whom the threat to the above "values" emanates. This enemy, too, can be of a concrete or of an abstract nature. It can be "the" Jews, Huns, Boches, tyrants, etc., or abstract concepts like world capitalism, Bolshevism, fascism, and any other kind of ism; it can be heresy, dogmatism, scientific fallacy, or what not. . . . A third factor contributing to the environmental situation eliciting the response is an inspiring leader figure. . . . A fourth, and perhaps the most important, prerequisite for the full eliciting of militant enthusiasm is the presence of many other individuals, all agitated by the same emotion.[99]

That Lorenz's specification of the four eliciting conditions of instinctual militant enthusiasm parallel the social conditions he and other members of his generation experienced during the Nazi era is striking. The four stimulus conditions correspond to, respectively, (1) the German *Volk*, threatened by the danger of biological annihilation by the (2) hateful (diseased, criminal, and biologically degenerate) Jew. The *Volk* will be protected by (3) the inspiring leader, the Führer, Hitler, who will (4) inflame the emotions of all members of the superior, Aryan race and thus elicit actions—militantly enthusiastic actions—aimed at destroying totally the arch, biological enemy of the *Volk*, the Jew.

It is a remarkable coincidence that Lorenz would be able to report, more than twenty years after the end of World War II, that there was scientific "evidence" for the existence of a reflex in humans that in effect freed the German people from any guilt in following Hitler. Indeed, it is bordering on the incredible that Lorenz "discovered" an instinctual reflex whose path of elicitation paralleled exactly the social events involved in Hitler's "war against the Jews."[100] We humans certainly cannot be guilty if we possess a knee-jerk reflex, since we were "designed" by evolution to possess such an automatic reaction. In the same sense, it would be consistent with Lorenz's argument to assert that the people of Nazi Germany could not help but follow Hitler once their "militant enthusiasm" reflex was imprinted and conditioned in the manner that occurred during the Third Reich. In other words, who could fairly blame the German people for the militant enthusiasm with which they murdered the Jews, if they were acting in the unthinking, irreversible, and reflexive manner that Lorenz said was the case with instinctual military enthusiasm?

But what if the notion of "instinct," as a hereditarily predetermined, genetically fixed and immutable set of behaviors, is a scientific fiction? What if the very behaviors that Lorenz described as genetically predetermined to emerge are neither inevitable nor immutable? What if there is no instinctual reflex such as militant enthusiasm—and thus no evolutionarily preprogrammed apologia for Nazi genocide? What if, even in the fish or the bird, much less the human being, nurture can alter both the nerve cells and the behaviors purportedly associated with instincts? In short, what if the supposedly predetermined and fixed genes / nerve cells / behavior connection is neither predetermined nor fixed but instead a readily modifiable, "plastic," linkage? Then social policies and programs designed to redirect innate militarism would be time and money misspent. The error in such policies and programs is that they are based on the assumption of an evil (or at least undesirable) basic nature for human beings, and if this assumption is wrong, then the programs that follow from it have no justification.

What can be wrong with promoting social policies and programs to diminish aggression and militarism? Even if those behaviors are not *really* instinctual, humans do engage in them all too often. Would the time and money spent on such programs then be wasted or unjustified? There are at least three reasons why they would. First, by building social programs to counter the occurrence of a scientific fiction, one is legitimating the use of what is in effect a lie in order to shape social policy.

Second, and as the comparative psychologist T. C. Schneirla points out in an evaluation of Lorenz's *On Aggression,* the efforts directed at deriving policies based on a scientific "lie" divert limited resources from scientifically supportable policies, which may actually be less pessimistic than the predetermined, instinctual views of Lorenz.[101] For instance, if nature and nurture are both equally involved in shaping human behavior, programs to develop positive and/or valued social behaviors may be designed proactively. There would be no need to have to expect *only* the worst and to have as the only option the design of "containment" or "rechanneling" programs to constrain the undesirable but inevitable behaviors.

Third, when one legitimates a scientific lie for use in shaping social policy, one is creating a potential for the lie to be used again in other policy areas. If humans are instinctually militaristic, might they also be controlled by other instincts? Can we not find instincts or, in other words, innate or inborn behavioral differences to account for differences between blacks and whites (for example, in intelligence), between men and women (for example, in their sexuality and family orientations), and between the socially privileged and the socially powerless (for example, in their resources and life options)? We certainly can, and, as colleagues of Schneirla pointed out, the biological determinist thinking exemplified by Lorenz has been used to legitimate not only militarism but also racism, sexism, and Social Darwinism.[102]

The view of human nature exemplified by Lorenz leads to a pessimistic, indeed bleak, view of our social world: humans have evolved to possess genes that inevitably give them specific behaviors. Some of these behaviors— for example, aggression—are shared by all people because evolution has provided all humans with an *almost* identical array of genes (i.e., an *almost* equivalent genotype). There are, however, some differences, and I will discuss the implications of these differences in Chapter 6. But to Lorenz it is those differences that are the basis of the most socially important (in my view, pernicious) implication of his concept of instinct.

Some differences are obvious—for instance, between men and women, females have two "X" chromosomes, and men have one "X" and one "Y" chromosome. Other differences may be more subtle and complex, reflecting the differing evolutionary histories of particular groups. In all cases, however, Lorenz, as an ardent Darwinist, would hold that genetic differences are outcomes of differences in the history of selection experienced by the groups in question.[103]

The social-policy implications of these genetic differences arise when, in Lorenz's view, the different selection histories involve civilization's attempts to domesticate and permit the continued survival of individuals who, under the conditions of natural selection, would not otherwise have survived.[104] This brings us back to the issue of "domestication-induced degeneracy," a theme of central concern to Lorenz during the Nazi era. Let us focus on this topic within Lorenz's post–World War II writings.

SELECTION AND ETHICAL DEGENERATION IN MODERN CIVILIZATION

In *Civilized Man's Eight Deadly Sins* (1974), Lorenz spells out the perils to modern human beings quite specifically and warns: "If progressive infantilism and growing juvenile delinquency are in fact, as I fear, symptoms of genetic deterioration, then we are in gravest peril. . . . Our environmental estimation of normal behavior is, through domestication, being destroyed or at least endangered."[105] Thus, to Lorenz in 1974, domestication has eroded humans' sense of the normal: their ability to tell the difference between pathological and nonpathological, or between good and healthy versus bad and unhealthy. Moral or ethical deterioration, then, is an outcome of domestication phenomena.

Is such ethical degeneration essentially the result of a generalized decline in the genes of all humans, or are there individual differences? Is the threat more a matter of some individuals carrying inferior genes—genes that produce moral and ethical degeneracy? Simply, did the post-Nazi-era Lorenz believe, as did the Nazi-era Lorenz, that some people have ethically inferior genes and some have ethically superior ones? Did the postwar Lorenz continue to believe that the "moral imbeciles" that threatened the health and survival of the *Volk* in 1940[106] also exist today and put in peril the survival of all nondegenerate humanity, if not the German *Volk*? Ultimately, the key question is: "What did Lorenz believe must be done to protect humanity against such a threat?" Did he make recommendations redolent of Nazi-era selections and eliminations, albeit perhaps not using the "unfortunate terminology" of that era? Did he once again call on the best among us—those with their instinctual ethics intact—to help create selection procedures aimed at restoring the natural order?

In my view, the answer to these last questions is yes. In 1975 Lorenz wrote:

Selection is and always has been the main creative and developing agent, from the molecular stage at the very beginnings of life up to the process of gaining knowledge by falsification of hypotheses. . . . By the very achievements of his mind, man has eliminated all those selecting factors which have *made* that mind. It is only to be expected that humaneness will presently begin to decay, culturally and genetically, and it is not surprising at all that the symptoms of decay become progressively more apparent on all sides. . . . The genetic "domestication" of civilized man is, I am convinced, progressing quite rapidly. Some cardinal symptoms which are present in most of our domestic animals are an increase in size and the hypertrophy of eating as well as of sexual activity. That all three of these symptoms have noticeably increased in man during the short span of my own life, is, to say the least, alarming. . . . Equally widespread is the quantitative increase of eating and sexual drive, accompanied in both cases by a loss of selectivity in releasing mechanisms. One has only to go to a beach where many urbanized people are bathing to note the rapidly increasing incidence of fat boys and young men or to look at a great modern illustrated paper in order to be confronted with both symptoms in a thoroughly alarming manner. . . . Of course, I do not know for sure that these symptoms are genetic, they may well be cultural, at least in part, but that does not matter much. Cultural development is analogous to genetical evolution in so many areas that the causal distinctions become immaterial as regards the phenomenon here under discussion, except that cultural processes are not less, but more dangerous because of their incomparably greater speed. . . . I am convinced that it is one of technocracy's most insidious stratagems to avoid all coercive methods and to rely on kind-seeming reinforcements alone. . . . I do not think that a healthy philosophy of values can develop without a sense not only of what is good but also of what is evil. It is my chief reproach against the ideology of the pseudodemocratic doctrine that it tends to eradicate, throughout our whole culture, the sense of values on which alone the future of humanity depends. . . . I do not believe that the death penalty or incarceration are able to prevent our genetic stock from decay; in fact there is nothing left in civilized society which could prevent retrograde evolution *except our nonrational sense of values,* which I still believe and hope can take a decisive hand in human evolution, both genetic and cultural. . . . There is such a thing as good and evil, there are decent guys and there are scoundrels and the difference between them is indubitably partly genetic. No living system can exist without elimination, however humanely it can be brought about and however much one tries not to make it appear as a punitive measure. . . . We *know* that evolution stops on its way upward and steps backward when creative selection ceases to operate. Man has eliminated all selective factors except his own nonrational sense of values. We must learn to rely on that.[107]

That passage underscores quite clearly Lorenz's continuing belief in the genetic basis of society in general, and in the hereditary determination of either the ethical value or the ethical worthlessness of people in particular. As did Haeckel, Lorenz saw both genetic and cultural evolution as essentially interchangeable, if not identical, processes that have a mystical, or at least nonrational, component which imbues only certain people with a proper sense of values. In addition, I believe that Lorenz saw in 1975, as he did in 1940, the need to rely on the moral responses of the people who have their innate ethical values intact to lead society. We must rely on them to bring civilization back to the path of healthy evolution from which it has been diverted—given the loss of natural selection processes and, instead, the institution of "kind-seeming reinforcements," a phrase reminiscent of Ploetz's characterization of the "misguided" humanitarian social programs that allowed the genetically unfit to survive and reproduce.[108]

Moreover, as did Haeckel and the Nazi racial hygienists, Lorenz called for the institution of "creative selection" procedures, which we may infer should be conducted by those with the best values. What these procedures should involve is not specified. However, I have noted that Lorenz criticized technocracy's avoidance of coercive methods and insisted on the need for elimination procedures beyond the death penalty and incarceration to ensure the continued existence of human beings. These views, frankly, are too consistent with his writings and those of other Nazi-era racial hygienists to lead one to conclude anything other than the continuity of a core theme.

Finally, although Lorenz was apparently flirting momentarily with environmental, or cultural, causation, he ultimately continued to take a hereditarian stance consistent with the biological determinism of the Nazi racial hygienists: even if one were to label society's problems as cultural and not biological in origin, with close analysis one would learn that any causal distinction becomes unimportant. Because of the ontological and material priority of biological processes over cultural processes, the latter ones can be reduced to the former; as such, therefore, the differences between good and bad people are, at the core, biological.

If there could be any remaining doubt that Lorenz continued to insist on the biological basis of morality and ethics and on the possibility of sorting people into good and bad groups, or ethically superior and inferior groups, on the basis of their environmentally immutable genetic inheritance, then some of his own statements should remove this doubt. Shortly after Lorenz was awarded the Nobel Prize, an article by a freelance writer living in Munich, Vic Cox, appeared in the March 1974 issue of *Human Behavior*. Entitled "A Prize for the Goose Father," it summarized Lorenz's career, contained excerpts from an interview with him, and discussed Lorenz's Nazi past and current thinking and work. As part of that article a passage from Lorenz's 1940 publication on "Domestication-Caused Disturbances in Species-Specific Behavior,"[109] which I discussed earlier, was quoted. In two places on page 19 of Cox's article the passage was quoted. In both places the quote was *incorrect,* containing a small typographical error in one word which, nevertheless, changed the meaning of the passage. The passage was misquoted to read that Lorenz called for "a more severe elimination of the ethnically inferior than has been done so far,"[110] although Lorenz actually said: "a more severe elimination of the ethically inferior than has been done so far."[111]

An extra *n* had been added to the word "ethical," to make the word "ethnical." Thus, we may infer that in 1940 Lorenz called for elimination—indeed, the "extermination"—of people who were ethically inferior by virtue of their genes, but that he did not call for the elimination of any particular ethnic groups that because of their "race" may have been carriers of inferior genes. My inference about Lorenz's meaning is supported by his own words. Lorenz wrote a letter to *Human Behavior* to correct the typographical error and to clarify the views the misquoted passage represented. The letter, appearing in the September 1974 issue, reads in its entirety as follows:[112]

> I thank you very much for the readiness to correct what was obviously more an error of the printer than of the editor. However, I beg you to realize that changing ethical into ethnical ("A Prize for the Goose Father," March 1974) makes me appear a rabid racist, which I never was. I never believed in any ethnical superiority or inferiority of any group of human beings, though I strongly hold that ethical inferiority of individuals due to heredity or to bad upbringing (lack of motherly love during the first year of life) is indeed a reality, which has to be taken seriously.
>
> I should highly appreciate it if you could include that in the intended correction.
>
> Prof. Dr. Konrad Lorenz
> Altenberg, Austria

Although Lorenz thus insisted he was not a racist, claiming that he had never believed there was a group of humans who by virtue of their *ethnical* heredity are inferior, he *did* believe in 1974, and in the Nazi period, that there was a group of humans who by virtue of their *ethical* heredity are inferior. It is this group—the moral imbeciles and dregs discussed in 1940—that should be eliminated, Lorenz believed. Such a fine conceptual distinction about who is and who is not to be the target of such "special treatment" (to use the Nazi euphemism for extermination) provided little comfort to the men, women, and children who were sent to the gas chambers and crematoria. In addition, Lorenz's conceptual distinction between ethnical and ethical has little historical validity, given the fact that Nazi Germany equated ethical degeneracy with membership in a particular ethnic group, Jews (see Chapter 2).

LORENZ AND CONTEMPORARY SOCIOBIOLOGY

The core message of the Lorenz of the post–World War II era is therefore not conceptually different from that of the Lorenz of the Nazi era. There is continuity in his views of the basic causes of both individual behavior and the social order: they are evolutionarily based, genetic causes. There is also continuity in his beliefs about the basis of social problems and about the threats to civilization: the erosion of natural selection; the reproduction therefore of hereditary moral inferiors who otherwise would have not survived; and the degeneracy in healthy instinctual patterns produced by the domestication of these inferiors. And ultimately there is continuity in the remedies Lorenz sees as requisite for saving civilization: we must rely on the nonrational ethical responses of the genetically/ethically superior among us and charge them with creating selection procedures to replace the eroded natural selection ones, and thereby eliminate the threat that domestication-induced ethical degeneracy poses to our healthy genetic stock.

Given what I regard as the quite evident Nazi-era/post-Nazi-era continuity in the core message of Konrad Lorenz, it is extremely puzzling that he received the world's most prestigious scientific award. Nevertheless, when other biological determinist positions were then promulgated, the broad scientific and societal legitimation of biological determinist ideology that the Nobel Prize provided could not help but impart to that ideology, for scientists and citizens alike, an aura of believability and the impression that this was on the cutting-edge advance of "normal science." This creation of a biological determinist *Zeitgeist* (spirit of the times) may have contributed in part to the broad scientific and social attention E. O. Wilson received in 1975 when he announced the presence of the new, synthetic discipline of sociobiology.[113]

The affinity between the work of Lorenz and that of contemporary sociobiologists is highlighted by the frequent citations of Lorenz's work in the current sociobiological literature, citations usually made approvingly and in the service of marshaling support for one or another sociobiological idea.[114] In addition, this reliance on Lorenz's work has led some contemporary sociobiologists to go out of their way to defend Lorenz against often unnamed critics and vaguely described criticism.

For instance, sociobiologist Melvin Konner, in his 1982 book, *The Tangled Wing*, praises one of Lorenz's Nazi-era papers ("Der Kumpan in der Umwelt des Vogels" [The Companion in the Bird's World, 1935]), reminds the reader that Lorenz was a Nobel Prize recipient, and then, in what is clearly a non sequitur, asserts that people are incorrect if they judge Lorenz only on his late (but uncited) popular writings or on the (undescribed) comments of his (unnamed) critics: "It is a magnificent paper, not only informative and convincing, but sweeping, incisive, beautiful. Reading it gives an impression very similar to that gained by reading Freud's early anatomical writings: that one has been very wrong to judge Lorenz only by his late popular writings, or worse, secondhand, by the opinions of his critics."[115]

Among the several problems with Konner's vague defense of Lorenz is the marked continuity between the topics, themes, and opinions found in the writings of Lorenz published in the Nazi era and in the post-Nazi era: there was little change in Lorenz's views about domestication-induced degeneracy, about selection, and about the genetic basis of moral worth and moral degeneracy. Simply, Konner proposes a division of Lorenz's work that close analysis of Lorenz's views across this span of time does not support. What Konner's remarks do suggest, however, is that perhaps sociobiologists may look approvingly on the core ideas of Lorenz because his ideas are similar to the ideas they themselves promote. In the next two chapters I consider these latter, sociobiological ideas in more detail and draw further parallels with the work of Lorenz. The focus will be on the role of African-Americans, of women, and of men envisioned within the sociobiological perspective.

4

SOCIOBIOLOGY, THE NEW BIOLOGICAL DETERMINISM

In a Darwinian sense the organism does not live for itself. Its primary function is not even to reproduce other organisms; it reproduces genes, and it serves as their temporary carrier. . . . The organism is only DNA's way of making more DNA. . . . Scientists and humanists should consider together the possibility that the time has come for ethics to be removed temporarily from the hands of philosophers and biologicized.
—*E. O. Wilson*, Sociobiology: The New Synthesis

In 1975 E. O. Wilson published *Sociobiology: The New Synthesis*, crystallizing the presence of a "new" scientific discipline and contending that sociobiology would be the "master" synthetic discipline, the field enveloping all of behavioral and social science.[1] As might be expected from the fact that there are differences in the views of scholars associated with sociobiological theory and research, some of Wilson's colleagues in the field of sociobiology disassociated themselves from the scientific hegemony Wilson claimed for this new discipline; for instance, sociobiologist Robin Dunbar pointed out that Wilson had "created the impression that sociobiology was on the verge of replacing most of the disciplines in the social and behavioural sciences" and that "this, of course, is arrant nonsense since sociobiology does not, of itself, deal with much of the subject matter of these disciplines."[2] However, Dunbar went on to defend the importance of Wilson's views for social and behavioral science: "Wilson was, none the less, right to emphasize the importance of sociobiology in relation to these disciplines. What it in fact does . . . is to provide a unifying umbrella under which these disciplines can interact on common ground."[3]

CORE CONCEPTS IN SOCIOBIOLOGY

And what is this "unifying umbrella" of sociobiology? It is, in Wilson's view, the ubiquity of genetic influence on all facets of individual behavior and social functioning. Indeed, Wilson's views are the ultimate in genetic reductionism: the complexity of all social behavior and, indeed, as he later claimed, all human culture,[4] can be reduced to a few simple principles of genetics. The key principle involved in this reduction is the idea of "gene reproduction." To Wilson, Richard Dawkins, and other sociobiologists, the essential, core purpose of human life is *only* to reproduce genes.[5] As sociobiologist Melvin Konner puts it, "A person is only a gene's way of making another gene."[6] Similarly, sociobiologist Dawkins sees humans as only "survival machines—robot vehicles blindly programmed to preserve the selfish molecules known as genes."[7]

According to Wilson, a human being therefore "does not live for itself"; his or her "primary function is not even to produce other organisms, it reproduces genes, and it serves as their temporary carrier."[8] From this view, we have evolved not to produce other people but only to replicate our particular complement of genes. Our life spans represent only a relatively short period within the vast temporal span of evolution, during which we provide a temporary "house," or transport, for the genes carried within us. Given this machine-like view of the "human as transport," it is clear why Dawkins sees humans merely as "lumbering robots [housing genes that] created us, body and mind; and their preservation is the ultimate rationale for our existence."[9]

Given this core "gene reproduction" principle, it is evident also why Dawkins sees genes as "selfish." In his book *The Selfish Gene* (1976), he posits that genes are "concerned" with nothing other than self-replication, with reproducing themselves over and over as many times as possible. So for Dawkins, humans are really just complex photocopying machines. Their mating rituals, their family relationships, and their cultural institutions are all inventions in the service of gene reproduction.[10]

Inclusive Fitness and Aggression

This core genetic principle of life leads to several other ideas about how genes influence individual behavior and the social world. First, the more copies of one's genes one can send out into the world, the more one is fulfilling the ultimate "intentions" of one's genes. Technically, in the terms of sociobiology, the more copies one makes, the more one's inclusive fitness is increased. This concept of inclusive fitness derives from the sociobiological view that natural selection is the essential vehicle for evolutionary change. As Social Darwinists and Nazi racial hygienists also stressed, not all genes are able to compete equally in the face of the fierce and rigorous challenges imposed by the natural environment. Only the most aggressive genotypes will succeed in this struggle for survival. Although all genes are attempting to reproduce themselves as much as possible, to include as many copies of themselves in the gene pool of the future, not all genotypes are fit enough to succeed. Only the most adaptive ones will prosper. And adaptiveness, in this case, means aggression.

As did Konrad Lorenz, who saw human aggression as both inevitable and inherent in the human genome—to the point of providing for innate "militant enthusiasm"—sociobiologists stress that genes make humans innately aggressive. To Lorenz, not only do the genes of humans make them aggressive, but the "highest form" of humans should be the most aggressive—the most militaristic—since such action reflects the presence of genes that have been the most successful in the struggles of natural selection. In turn, the sociobiological world of selfish genes and the robotic "survival machines" that house them sees aggression as playing an even more central role in the survival and propagation of genes. Aggression functions to allow genes to enhance their inclusive fitness; "blind" (that is, unthinking, machine-like) aggression allows genes to eliminate anything in the environment that interferes with their reproduction.[11] Thus, in the sociobiological conception "aggressiveness" equals "adaptiveness."

To Lorenz and to sociobiologists, selfish human aggression is the cornerstone of the control genes have over human functioning. This conceptualization of the genetic basis of aggression supports the connection between the genetic determinism of Lorenz, and of Nazi ideologues, and the genetic determinism of sociobiologists. In both groups the most successful genes, the "Best" of evolution, will be the most successful at ruthless aggression; they will be the most selfishly directed to maximizing their presence in the gene pool and, at the same time, to minimizing the presence of other genes. Of this blind, ruthless, militant aggression, Dawkins says:

> To a survival machine, another survival machine (which is not its own child or another close relative) is a part of its environment, like a rock or a river or a lump of food. It is something that gets in the way, or something that can be exploited. It differs from a rock or a river in one important respect: it is inclined to hit back. This is because it too is a machine which holds its immortal genes in trust for the future, and it too will stop at nothing to preserve them. Natural selection favours genes which control their survival machines

in such a way that they make the best use of their environment. This includes making the best use of other survival machines, both of the same and of different species.[12]

Similarly, and redolent of Lorenz's views in his *On Aggression*, Melvin Konner states: "I believe in the existence of innate aggressive tendencies in humans."[13] He also argues: "If we are ever to control human violence we must first appreciate that humans have a natural, biological tendency to react violently, as individuals or as groups, in certain situations."[14]

We have seen that, to Lorenz, these innate violent reactions are elicited in a reflex-like manner among either individuals or groups; the reflex occurs when members of the in-group (to Lorenz, the *Volk*) are threatened by members of the out-group (to Lorenz, the dregs of society, the human "cancers," who in Nazi Germany were most notably the Jews). Lorenz believes aggression is released on the basis of genetic/racial in-group membership versus out-group membership, and indeed, as one scholar puts it, he "regards the formation of opposite terms, the juxtaposition of Alpha and Nonalpha, as a way of thinking that is innate."[15] Similarly, we have seen that Dawkins contends that selfish genes impel humans to act aggressively toward survival machines not of their own, close genetic group (i.e., a child or another close relative), and E. O. Wilson also believes that fear of (and hatred toward) an out-group—xenophobia—is innate in humans.[16] For example, in his book *On Human Nature*, Wilson wrote:

> Culture elaborates the rites of passage—initiation, marriage, confirmation, and inauguration—in ways perhaps affected by still hidden biological prime movers. In all periods of life there is an equally powerful urge to dichotomize, to classify other human beings into two artificially sharpened categories. We seem able to be fully comfortable only when the remainder of humanity can be labeled as members versus nonmembers, kin versus nonkin, friend versus foe.[17]

Thus, to Lorenz and such sociobiologists as Dawkins, Konner, and Wilson the genes of humans make them innately and ruthlessly aggressive, a characteristic of behavior that is directed especially at people not of their own group. The inherent struggle, or *Kampf*, of natural selection requires that a genotype be completely selfish in its attempts to reproduce as many copies of itself as possible. Because other genotypes have the same goal, only the most aggressive genotypes will be able to exploit effectively the limited resources of the environment and to reproduce themselves maximally.

Therefore, in the views of Lorenz and the above-noted sociobiologists, fear and hatred of an out-group, aggression, ruthlessness, and selfish exploitation of limited environmental resources are conflated to define the most successful, the most fit, the evolutionarily "Best" human genotypes. In other words, because they exist in a world of severe challenges and limited resources, all genotypes must struggle arduously to include as many copies of themselves in the gene pool as possible. The more replicates of one's genes there are in the gene pool—that is, the greater one's inclusive fitness— the more fit for survival, the more adaptive, those genotypes are. That is the case, of course, *if and only if* the rigors of natural selection have not been distorted or eliminated by societally constructed institutions that afford less-fit genotypes the opportunity to survive and reproduce.

Gametic Potential and Gender Differences in Parental Investment

Another genetic principle promoted by sociobiologists derives from the striving of all genotypes to maximize their inclusive fitness. Termed *gametic potential,* this principle has important social-policy implications for socio-biologists, particularly for the roles of men and women in society. Just as all genotypes are not equally fit for having their replicates in the pool of genes, so too do men and women differ in their potential for transmitting copies of their genes into the future. For example, Konner claims that "as now seems clearly demonstrated, there are biological reasons why women, like other primate females, have a weaker aggressive tendency than males."[18] Because, according to sociobiologists, aggression is the key to getting one's genotype reproduced maximally, and because women lack an aggressive ability sufficient to compete with men, women must evolve some other strategy to enhance their inclusive fitness. The strategy derives from the nature of the specialized cells women use to transmit copies of their genes to future generations.

Men's and women's genotype copies are contained in their respective sex cells, or gametes—that is, their sperm and ova. Both types of gametes function to maximize the inclusive fitness of the genotypes they carry. However, the two types of gametes have a different *potential* for such reproduction—due to the anatomical and physiological differences between the "lumbering robots" (men and women) housing these gametes. Luckily for men—who (it is of more than passing interest to note) were the founders and leading proponents of sociobiology—many genotype copies can be made. Their gametes can be "sent forth to multiply" quite readily; millions can be sent out with each ejaculation. Thus, in the terms of sociobiology, their "gametic potential" is great, given that there is at least theoretically a large and ready pool of recipients of their gametes. Sociobiologist Daniel Freedman puts it this way: "Since mammalian males produce many more sperm than females produce ova, any given male has far greater potential for producing offspring. He is also more inclined to compete with other males over the 'scarce' resource, females."[19]

Any male therefore has a greater potential for enhancing his inclusive fitness than any female, given males' greater gametic potential. Moreover, males *must* have in general a more aggressive genotype than females, since they must compete for access to the female gamete, viewed as a "resource" for the deposit of the male's sperm. Such competition is highly desirable, in the view of sociobiologists, because it ensures that the most aggressive genotypes—those best suited to succeed against the struggles of natural selection—will reproduce most often. Indeed, sociobiologist Robert Trivers has argued that a major basis for war is the genetically based need for men to reproduce their genes by impregnating women of the defeated enemy. He contends that it is a "fact that warfare can increase the reproductive success, that is, the surviving offspring of the victors in the war" and continues:

> One of the most striking characteristics of warfare, and certainly classical warfare, is that when you overrun the other country you loot and pillage but you also grab up the women and you either inseminate them on the spot or you take them back as concubines. You kill off the adult males; you sometimes castrate young boys and bring them back as servants. So I think warfare has traditionally had a strong sexual counterpart to it which is certainly biological and you do not have to look far to see that there is that tendency running today.[20]

In short, the ideal successful male genotype is an aggressive one, capable of conquering the rigors and challenges of the world and garnering its resources, one of which is a share of the ova available for fertilization by these aggressive genes. The aggressive male genotypes impel the lumbering robots housing them to reproduce with as many females as possible. In turn, the genotypes of females impel women to try to reproduce in quite another way.

One might understand the origin of this "alternative," female reproductive strategy if one asks: "Given the vast difference in reproductive potential, and if the point of life is to actualize such potential, is it not reasonable to expect that on the average the male pattern of courtship will differ from the female? Might nature not have arranged it so that men are ready to fecundate almost any female and that selectivity of mates has become the female prerogative?"[21] In answer to such questions, van den Berghe and Barash note: "Human females, as good mammals who produce a few, costly and therefore precious, offspring, are choosy about picking mates who will contribute maximally to their offspring's fitness, whereas males, whose production of offspring is virtually unlimited, are much less picky."[22]

Implications for Social Behavior and Development

What, then, do the different gametic potentials of the sexes imply for an understanding of the social behavior of men and women? Given the selfishness of genes, and their single-minded direction of the photocopying machines housing them, men have developed sexual mores involving the acceptability (if not the appropriateness) of having multiple sexual partners. Indeed, Pierre van den Berghe and David Barash argue that the different gametic potential of men and women explains "the widespread occurrence in human societies of polygamy, hypergamy, and double standards of sexual morality":

> There is another related reason for the sexual double standard in such things as differential valuation of male and female virginity and differential condemnation of adultery: marital infidelity of the spouse can potentially reduce the fitness of the husband more than that of the wife. Women stand to lose much less if their husbands have children out of wedlock than vice versa. . . . In addition, a woman will, at a maximum, produce some 400 fertile eggs in her lifetime, of which a dozen at most will grow up to reproductive age, while a man produces millions of sperm a day and can theoretically sire hundreds of children. Not surprisingly, females tend to go for quality, and males for quantity. This means, for women, to select as mates those men most likely to contribute to the fitness of their children, both in terms of physical and social attributes. . . . Men have power, and hence control over resources. Women, as do the females of many other species, choose males with the best possible resources.[23]

Moreover, given the large number of offspring he has the potential to produce, a male's *parental investment* in any one is quite small. In other words, men should not invest themselves appreciably in the care of any particular one of their children, and this cavalier attitude toward childrearing is predicated on the fact that any given child or set of their children is but a small sample of what they could produce.

Unfortunately for the recipients of males' genetic copies—women—their gametic potential is quite different, and so is their parental investment. At best, they can replicate themselves only every nine months. Even in the case of multiple births (twins or triplets), a woman could not ever replicate her genes as much in a lifetime as a man could do in only a week or two. Therefore, a woman's investment in her offspring is much greater than a man's. Moreover, because they cannot reproduce very frequently, women will not be motivated toward frequent copulation with multiple partners. Instead, women's need to protect their offspring and ensure their survival should motivate them to keep their impregnators bound to them, and consequently the female of the species develops monogamous sexual behaviors and a devotion to childbearing and rearing. Van den Berghe and Barash argue:

> For a woman, the successful raising of a single infant is essentially close to a full-time occupation for a couple of years, and continues to claim much attention and energy for several more years. For a man, it often means only a minor additional burden. To a limited extent, sexual roles can be modified in the direction of equalization of parental load, but even the most "liberated" husband cannot share pregnancy with his wife. In any case, most societies make no attempt to equalize parental care; they leave women holding the babies.[24]

And, lest anyone contend that these different moral, sexual, and social orientations of men and women are merely products of socialization, sociobiologist David Barash argues that the sex differences in gene reproduction strategy explain "why women have almost universally found themselves relegated to the nursery, while men derive the greatest satisfaction from their jobs."[25] Similarly, van den Berghe and Barash note that "ethnographic evidence points to different reproductive strategies on the part of men and women, and to a remarkable consistency in the institutionalized means of accommodating these biological predispositions" and therefore conclude: "Men are selected for engaging in male-male competition over resources appropriate to reproductive success, and . . . women are selected for preferring men who are successful in that endeavor. Any genetically influenced tendencies in these directions will necessarily be favored by natural selection."[26]

Dawkins embellishes this idea that the different gametic potentials of women and men account for differences in their personal and social behavior. He contends that women's *exploitation* by men is biologically determined. In his view, women's exploitation is evolutionarily bred, genetically inevitable, and simply the natural way of things. Accordingly, Dawkins argues that it is not only the different *number* of sex cells that can be used for genotype reproduction which differentiates the behavior of the sexes, but also the different *size* of the sex cells:

There is one fundamental feature of the sexes which can be used to label males as males, and females as females, throughout animals and plants. That is that the sex cells or "gametes" of males are much smaller and more numerous than the gametes of females . . . it is possible to interpret all the other differences between the sexes as stemming from this one basic difference. . . . Sperms and eggs . . . contribute equal numbers of genes, but eggs contribute far more in the way of food reserves: indeed sperms make no contribution at all, and are simply concerned with transporting their genes as fast as possible to an egg. At the moment of conception, therefore, the father has invested less than his fair share (i.e., 50 per cent) of resources in the offspring. Since each sperm is so tiny, a male can afford to make many millions of them every day. This means he is potentially able to beget a very large number of children in a very short period of time, using different females. This is only possible because each new embryo is endowed with adequate food by the mother in each case. This therefore places a limit on the number of children a female can have, but the number of children a male can have is virtually unlimited. Female exploitation begins here. . . . Each individual wants as many surviving children as possible. The less he or she is obliged to invest in any one of those children, the more children he or she can have. The obvious way to achieve this desirable state of affairs is to induce your sexual partner to invest more than his or her fair share of resources in each child, leaving you free to have other children with other partners. This would be a desirable strategy for either sex, but it is more difficult for the female to achieve. Since she starts by investing more than the male, in the form of her large, food-rich egg, a mother is already at the moment of conception "committed" to each child more deeply than the father is. She stands to lose more if the child dies than the father does. More to the point, she would have to invest more than the father *in the future* in order to bring a new substitute child up to the same level of development. If she tried the tactic of leaving the father holding the baby, while she went off with another male, the father might, at relatively small cost to himself, retaliate by abandoning the baby too. Therefore, at least in the early stages of child development, if any abandoning is going to be done, it is likely to be the father who abandons the mother rather than the other way around. Similarly, females can be expected to invest more in children than males, not only at the outset, but throughout development. . . . The female sex is exploited, and the fundamental evolutionary basis for the exploitation is the fact that eggs are larger than sperms.[27]

Thus, natural selection has shaped men and women to behave—morally, sexually, and socially—in quite distinct manners. Evolution has created sex differences in gametic size and number that provide for the genetic determinism of significant contrasts in the behavior of men and women, especially of the biologically based and hence inevitable exploitation of women by men. Such differences are not only *"in"* our genes"; they are also in effect our genes' strategies for their own reproduction. Extramarital affairs provide one illustration of this genetic control over human moral and sexual behaviors. The different genetic potentials of males and females "legitimates" extramarital sex for males but not for females. And not only is male promiscuity "excused" because of differential genetic potential, but so too is the use of violence toward wives who are having extramarital sexual relations. To explain these sex differences, we may note van den Berghe and Barash's observation:

> It is true, of course, that social advantages of wealth, power, or rank need not, indeed often do not, coincide with physical superiority. Women in all societies have found a way of resolving this dilemma by marrying wealthy and powerful men while taking young and attractive ones as lovers: the object of the game is to have the husband assume parental obligations for the lover's children. Understandably, men in most societies do not take kindly to such female strategies on

the part of their wives, though they are not averse to philandering with other men's wives. The solution to this moral dilemma is the double standard.[28]

Expanding on this point, Freedman notes:

All things considered, however, it would appear that men are indeed more polygamous in their motivation than women, so that it behooves women to become expert in selecting males who will not desert them and their children. . . . Cultures, after all, did not appear out of the blue . . . and we have to assume that cultural universals reflect those aspects of our species that were evolutionarily derived (evolved). Male promiscuity is universally winked at because there is nothing much we can do about it, and Kinsey's main findings appear to be descriptions of the species: males must have frequent "outlets" for sex, whether heterosexual or homosexual; whereas many females can go for long periods without copulation or masturbation. . . . And this difference appears to hinge on the difference in gametic potential that we have been discussing. . . . I think we can safely make the following generalization: the problem for the mated female . . . is to prevent her mate's straying; if she cannot control his sexual activity, it is most important to prevent his becoming attached to another female, specifically a female she cannot dominate. . . . Thus, in humans, relations with prostitutes are generally tolerated for an attachment to a whore is everywhere deemed unlikely. As in the gelada baboon, in humans female jealousy is based not on the male's sex act with another woman but on his potential attachment to the latter. . . . Male jealousy is rather different. It is one thing to have (and impregnate) many women, but even the sated sultan makes sure that there are no sexual intruders. Probably all men . . . fear cuckoldry. . . . The evolutionary point of male jealousy is obvious. It does not make evolutionary sense for the male to invest in a child not possessing his genes, and the murderous jealousy exhibited by a cuckolded male is biologically sensible. Such a response, we can surmise, evolved to minimize female cheating and to assure that the children a man supports are his own. Furthermore, the cuckold's retribution can strike either the female or the male cheater, so that both cheat only with immense trepidation. Even a weakling husband is dangerous when cuckolded, and most legal systems (perhaps *all* patrilineal systems) wink at the ensuing violence.[29]

Dawkins extends across the life span this story about the biological basis of men's promiscuous (for example, extramarital) sexual interests. He offers both "a possible explanation of the evolution of the menopause in females" and an account of the sociobiological basis of the existence of what are colloquially (and pejoratively) termed "dirty old men": "The reason why the fertility of males tails off gradually rather than abruptly is probably that males do not invest so much as females in each individual child anyway. Provided he can sire children by young women, it will always pay even a very old man to invest in children rather than in grandchildren."[30]

In sum, sociobiological ideas forwarded by Freedman, Dawkins, van den Berghe, Barash, and others indicate that it is in the nature of the world for men and women to differ significantly in their sexual and social behaviors.[31] Men—impelled mechanistically by their genes—are driven to seek sex with as many women as possible, to achieve more and more copies of their genes and to be not overly devoted to or concerned with any one or any few given replicates. Women, on the other hand, must remain monogamous homebodies in order to maximize the probability that their relatively few replicates will survive. In essence, then, men and women are genetically impelled to differ in ways that are consistent with traditional—that is, stereotypic—sex-role patterns. Psychologist and sociobiological writer S. K. Kachigan puts it this way:

> The *limited number of desirable males* vis-à-vis the number of desirable females, is entirely consistent with the theory that *polygamy is the natural order among humans,* as it is with most species in the animal world, representing a *genetic* dynamic that promotes the optimum composition and viability of successive generations. . . . This gender dynamic implies that females are *competing* with one another for a *limited number of choice males,* and in its strength and consistency represents a *behavioral manifestation of an inborn genetic tendency.* This in conjunction with the implied disparity in the number of desirable males and females in the population, strongly suggests that *polygamy is the natural biological order among humans.*[32]

The sociobiological views promoted by Wilson in effect involve the claim that "anatomy is destiny," and the same is true of Sigmund Freud and later

Erik Erikson.[33] But the views of Wilson, Dawkins, Freedman, and other sociobiologists, such as van den Berghe and Barash—that genetically predetermined sex-role differences are both inevitable and necessary for human survival—are cast in the parlance of abstract mathematical models of evolutionary change. Darwinian ideas about evolution, natural selection, and species survival are then quantitatively modeled[34] and coupled with detailed and often elegant research with invertebrates (e.g., insects) and birds.

Wilson, Dawkins, Freedman, and other sociobiologists have built a nature edifice encompassing the very core of all human existence: the reproduction of men and women, the character of the family, and the survival of the species. If genes actually work the way sociobiology requires, any notions or nurture or of a nature-nurture fusion as a source of key features of human behavior (that is, features based in evolution and affording human survival) are mere fictions. Genes act as selfish, goal-directed, intentional agents, and after other, more superficial "causes" for our behavior are stripped away (for instance, "causes" involved in an individual's development, such as his or her learning or personality) genes provide the ultimate basis for our functioning, the replication of our genotype. In this ultimate sense, our social world does not interact with our genes, much less act as an alternative source for our behavior. Instead, to sociobiologists, our social world—our mores (for example, regarding sexual permissiveness or monogamy), our social institutions (such as marriage and the family), and indeed our entire culture is nothing other than the outcomes of strategies laid down by our genes for their own replicaton.

Sociobiologists have complete faith in the inevitable reducibility of human behavior to the functioning of selfish genes, and, akin to Konrad Lorenz, this genetic determinism view has necessarily xenophobic and ruthlessly selfish implications for society. The faith in genetic determinism maintained by sociobiologists is expressed by Dawkins when he says, "It can be perfectly proper to speak of 'a gene for behavior so-and-so' even if we haven't the faintest idea of the chemical chain of embryonic causes leading from gene to behavior," and, "Be warned that if you wish, as I do, to build a society in which individuals cooperate generously and unselfishly towards a common good, you can expect little help from biological nature."[35]

To what extent is this sociobiological view of human behavior and of society supported in the scientific literature? What scientific evidence do sociobiologists use to legitimate their claims, and how adequate is this evidence? Let us turn to these key questions.

EVALUATING SOCIOBIOLOGICAL "EVIDENCE"

We have noted already that in 1975 E. O. Wilson defined sociobiology as "the systematic study of the biological basis of all social behavior,"[36] so it may seem surprising and a bit contradictory to learn that in 1980, in response to some critics, Wilson stated: "Contrary to an impression still widespread among social scientists, sociobiology is not the theory that human behavior has a genetic basis."[37]

Perhaps Wilson was just playing with words. Perhaps he meant that sociobiology is not a "theory" but only a "perspective," or merely a rather general framework within which to study systematically the biological and therefore ultimately genetic basis of all social behavior. In any case, the issue is *not* whether Wilson is hinging his claims about the mistaken impressions social scientists have about sociobiology on the difference in meaning between the phrases "the theory that" and "the systematic study of." The issue is that Wilson's own words show that sociobiology *is* the study of the role of the connection between genes and human social behavior. In fact, he uses the term "sociobiological theory" to represent this linkage. He claims:

Real sociobiological theory allows no less than three possibilities concerning the present status of human social behavior:

(a) During the rapid evolution of the human brain, natural selection exhausted the genetic variability of the species affecting social behavior, so that today virtually all human beings are identical with respect to behavioral potential. In addition, the brain has been "freed" from these genes in the sense that all outcomes are determined by culture. The genes, in other words, merely prescribe the capacity for culture. *Or,*

(b) Genetic variability has been exhausted, as in (a). But the resulting uniform genotype predisposes psychological development toward certain outcomes as opposed to others. In an ethological sense, species-specific human traits exist and, as in animal repertories, they have a genetic foundation. *Or,*

(c) Genetic variability still exists, and, as in (b), at least some human behavioral traits have a genetic foundation.

Having identified these alternatives, and stressed the freedom of the discipline of sociobiology from the necessity of any particular outcome, I can now add that the evidence appears to lean heavily in favor of alternative (c).[38]

In the case of each alternative—(a), (b), and (c)—the emphasis is on the link between evolution, genetic variability, and human behavior and society. However, if sociobiology has spent a good deal of time exploring the first two of the three alternatives, such work has not found its way into the published literature—that is, Wilson is correct in asserting that to the extent that "evidence" does exist in support of any of the three alternatives it does so in regard to alternative (c). Yet to the extent that this support does exist, it is not because all three alternatives have been repeatedly subjected to comparative scientific analyses. Rather, the preponderance of published sociobiological work—at least insofar as the human literature is concerned—has taken as its "working assumption" the idea found in alternative (c). The "evidence" derived from such work constitutes not a test of competing "hypotheses" but an attempt to bring empirical observations to bear on a demonstration of a guiding presupposition.

Given what are quite well known facts of genetic variability, it would be preposterous to conduct a scientific investigation predicated on the idea that genetic variability does *not* exist. There is abundant evidence that truly vast human genetic variability does exist (see Chapter 6). Certainly Wilson and other sociobiologists are fully aware of this quite basic evidence, so it seems equally certain that sociobiologists would pay no serious scientific attention to alternatives (a) and (b). Therefore, those two alternatives cannot be and are not treated as viable counters to alternative (c), and the last conception is the only one actually pursued scientifically by sociobiologists. But, given that no alternatives are really comparatively tested, such pursuance is more in the direction of *demonstrating* how empirical phenomena coincide with a conceptual presupposition than of conducting a critical test of theoretical alternatives.

In such demonstrations, three types of evidence have been used. It is most useful to discuss and evaluate each type of evidence separately.

The Concept of Homology

Sociobiologists demonstrate that human social behavior is constrained by evolutionarily shaped genes by drawing parallels between the behaviors of humans and of nonhuman animals. If the behaviors of the two species can be described similarly, then (it is argued) there must be some evolutionary connection, or some continuity, between them. A common evolutionary pathway for a physical structure or a behavioral function in two species is termed a "homology." Sociobiologists argue that when the characteristics of two species can be described in a common way, it is evidence of homologous evolution, and the positing of such homology is presented as proof that the characteristics in question are controlled, or constrained, by evolutionarily shaped genes.

The writings of sociobiologist Daniel Freedman illustrate how such "evidence" can be used. In attempting to document his views that the gametic potential of human males makes them show sexually promiscuous behavior—in order to increase their opportunities to garner the "scarce resource" of females' ova—while human females' gametic potential makes them more monogamous, Freedman finds homologies between fruit flies, rhesus monkeys, and jungle-dwelling humans in South America:

> Bateman has . . . worked with fruit flies and has noted some important trends. In a carefully controlled study, he found that whereas almost all females have young, a small proportion of the males in a population do a large proportion of the successful mating. Whereas only 4% of the females produced no progeny, 21% of the males produced none. It is perhaps astounding that we have similar figures among mammals. Koford, working with free-ranging rhesus monkeys imported to Cayo Santiago island (just off Puerto Rico's coast), observed: "Roughly, the highest fifth of the males performed four-fifths of the observed copulations. . . . At least half of the males of breeding age [copulated] rarely or not at all, apparently because of social inhibition."
>
> It is probably more than coincidence that similar rates of male mating success are found among the polygynous Yanomamo in the Venezuelan and Brazilian jungles. . . . About the same proportions of males have the majority of wives and children: in any village some males will have 4 or more wives and 20 or more children, others rather fewer wives and children; but about one-fourth of mature males will be childless. Chagnon reports that the variance in numbers of offspring among men of a village can be as much as 10 times that of females. That is to say, among men, the variability from man to man to man is as much as 10 times greater than the variability from woman to woman. . . . Thus, in fruit flies, rhesus monkeys, and some polygynous tribes, females tend to cluster about an average number of young whereas males form a greatly skewed curve, some very successful, many not successful at all. And, since most mammals are polygynous . . . this tendency may characterize the entire class Mammalia.[39]

Freedman carries his argument one step further. By again using what he regards as common behavioral descriptions across species, he attempts to provide an evolutionary and genetic account not only for inevitable human male promiscuity but also for the genetically preordained urge to seek sex with other females even to the point of forcing oneself onto them—that is, committing rape.[40] First, Freedman cites the work of Grzimek:

> In spring, when the gonads are at the peak of their development, there are attempts to "rape" strange females in the mallard and pintail and a few other species. Heinroth describes these raping flights: "In a shallow bay of the pond the female of the mallard, pair A, seeks food by upending; her mate is close by keeping watch. A hundred feet away a second pair B, comes down on the water. Male A quickly swims towards the strange female B, finally flying towards her, but she takes flight at the last moment and a wild aerial chase begins. The pursued duck rises higher and higher, the strange drake A behind her; she tries to escape by swerving and suddenly flying slowly, and both are followed by male B since he does not know where his mate will end up. Thus one sees two drakes following a duck, and this is generally interpreted as if both were chasing the duck; in reality, however, a strange male is chasing the female of a pair which belong together. Gradually male A gives up and returns directly to his mate.[41]

Then Freedman makes an inference about the "promiscuous, polygynous intentions" of ducks and draws a conclusion about the insatiable, continuous, and carnal search by human males for females with which to copulate:

> It would appear that if the mallard drake had his way his would be a polygynous species and, in fact, one does occasionally see a consortship of two females and a male.
> In our own species and our own culture, I am asserting nothing startling when I point out that with sexual maturity, most heterosexual males are in constant search of females, and if inhibited about sexual contact, they fantasize almost continuously and fairly indiscriminately about such contact. Like Philip Roth's *Portnoy*, adolescent males in our culture frequently experience life as a nearly continuous erection—spaced by valleys of depression that accompany sexual disappointment.[42]

Are the above-noted descriptions, and those of other sociobiologists, about purportedly comparable human and nonhuman social behavior,[43] satisfactory proof of the evolutionary and genetic bases of human behavior? Does apparent descriptive similarity establish evolutionary homology? The answer to both of those questions is no. There are several reasons for this answer, not the least of which is how difficult it is to accumulate sound scientific evidence showing common evolutionary descent, even when physical attributes are being considered.[44] This task is even more difficult with regard to behavioral characteristics; even very similar behaviors may (1) have quite different underlying processes and/or (2) have quite different functions.[45]

In regard to (1), it is true that one can describe similar behaviors across even very different species. For instance, insects, fish, rats, and humans all *learn*—that is, in members of each of those species systematic and relatively permanent changes in behavior occur in relation to experience. Nevertheless, the ways in which those species learn, the processes of learning, vary considerably. For example, it would be difficult to contend that thought processes—cognition—play a part in the learning of insects at any point in their lives. In turn, it would be equally difficult to argue that cognition does not enter into human learning for anything other than the earliest portion of the life span (and even there cognition may still play a role).[46]

Of course, any animal can change its behavior in the face of environmental alterations, so in this sense any animal can be said to be capable of learning. However, what is involved in learning within different animals is by no means the same. Experience-based changes in behavior may allow all animals to adjust to the current features of their environment, but that similarity is at best evidence for an analogy, not for a homology.[47] Animals have different processes, which may subserve analogous functions. Insects may have reflex-like processes that allow them to adjust to changed environmental conditions; humans may have complex and plastic cognitive processes that allow them to fit well with their environmental circumstances. Although such distinct processes serve analogous purposes in that both allow adjustment to a changed environment, a claim that such descriptive analogies are indicative of common evolutionary histories is at best egregiously naive and at worst poor scholarship. Robin Dunbar is frank in admitting this limitation in sociobiological scholarship: "Many of those who were influential in promoting the sociobiological perspective . . . (e.g., E. O. Wilson) . . . tend to be unaware of the more sophisticated nature of the behaviour of higher organisms and are apt to regard even advanced mammals simply as scaled-up insects."[48]

In regard to point (2), the presence of two identical behaviors in different organisms does not constitute proof for even common function or purpose. To illustrate, the reasons male mallards might force copulation on a female of their species are certainly distinct from those involved when a human male forcibly copulates with a human female. To label both the male duck's behavior and the actions of the human male with the identical term— "rape"—seems to trivialize, through biological reductionism, what is certainly a complex and violent human act that, current scholars point out, may not even be a behavior predicated in any way on sexuality or sexual feelings.[49]

Freedman, Barash, and other sociobiologists argue for homology on the basis of such cross-species descriptions. Can they contend, however, that the devaluation of women in many sectors of modern society, and the legitimation of violence as a means of exercising social (and political) control, do not enter into the primary causation of forced copulation among human males and/or do enter into the basis of such behaviors in ducks? I think not. The mere superficial portrayal of behaviors in two species as appearing comparable is no proof at all of their common evolutionary heritage, no proof of the extent to which such behaviors are genetically constrained or produced.

Indeed, and underscoring again the differences among sociobiologists on particular issues,[50] Wilson too seems to have come to that conclusion: "We cannot rest the hypothesis of genetic constraint in human social behavior on the indirect evidences of homology."[51] If the sociobiologists' type of behavioral homologies does not constitute adequate proof for the genetic basis of human social behavior, what does? Two other types of evidence have been offered: one pertaining to the concept of the heritability of human behavior, the other pertaining to the adaptive character of human behavior.

The Concept of Heritability

Wilson has argued that the third of the three possible theoretical alternatives on which sociobiology rests is the one that current evidence favors heavily: that genetic variability exists and that at least some human behavioral traits have a genetic foundation.[52] However, this seemingly straightforward statement leads to a thicket of conceptual confusion.

Sociobiology focuses on behavioral characteristics, or traits, that are *common* to the species. Males, having their particular gametic potential, show a specific sexual orientation (promiscuity and an instinctual urge to copulate with multiple females); females, having their particular gametic potential, show their own specific set of sexual and social behaviors (monogamy and a child- and family-centered institutional orientation). The task of the sociobiologist is to show scientifically that such traits uniformly and unequivocally characterize the subgroup of humans in question (males and females, in this example) because they possess evolutionarily based genetic "directives" for genotype reproduction. The sociobiologist's goal is to demonstrate that some human traits—specifically, common traits for a given group, having to do with that group's reproductive strategy and hence their inclusive fitness—have a genetic basis.

To demonstrate the common, or invariant, inheritance of these traits, sociobiologists often rely on the concept of "heritability." The argument is that if the trait in question is shown to be heritable, then the trait must be commonly inherited. Although, arguing that if a trait is heritable it is therefore inherited seems so obvious that it borders on tautology, nothing could be further from the truth. The demonstration of heritability says virtually nothing about the extent to which a trait is commonly inherited. In fact, evidence of heritability cannot be used as evidence for the common possession of a particular set of genes. Indeed, quite the opposite is the case. Heritability, I will explain, is a statistical estimate of genetic *variability* among people; it is an estimate of variation in genes, not of commonality. In fact, I will show that if the basis for a human trait was genes common to all people in a group (as sociobiologists need to establish), then the heritability for that trait would be zero.

If this preamble to a discussion of heritability gives the idea that sociobiologists are using the concept in a confused and confusing manner, that is precisely the case. One key reason for this confusion is that "heritability" itself is a difficult concept. At first blush "heritability" would seem to pertain to the extent to which something is inherited—that is, based in the genes. For instance, if we were told that "intelligence is 80 percent heritable,"[53] it might be reasonable to take this to mean that 80 percent of intelligence was genetically "determined"—that is, that any given person's intelligence was largely (80 percent) shaped by his or her genes and that something else—environment—shaped the small percentage remaining.

This reasonable interpretation, which scientists, members of the media, and government policymakers often make, is completely incorrect. Technically, heritability pertains only to differences *between* people; it has absolutely nothing to do with the extent to which anything—be it genes or environment—determines characteristics *within* an individual. What heritability does refer to is the extent to which differences between people in a specific characteristic can be described by genetic differences between these people, but even this reference to genetic differences is quite misleading.

Suppose there was a society that had strict gender discrimination laws and that one of those laws pertained to whether someone could be elected to government office. The law simply read that men could be elected to such positions of public leadership and that women could not be. What would a researcher need to know in order to divide completely correctly a randomly chosen group of people from this society into one of two groups? Group 1 would be made up of societal members who had a greater than 0 percent chance of being elected to a leadership post, and Group 2 would be those who had absolutely no chance. To make this division with complete accuracy, a researcher would need to know only whether the people had either an XX pair of chromosomes or an XY pair. If a person possessed an XX pair of chromosomes, the person would be a female, because possession of the XX chromosome pair leads to female development; if a person had an XY pair of chromosomes, that person would be a male. The researcher could correctly place all possessors of the XY pair into the "greater than zero chance" group and all possessors of the XX pair into the "no chance" group.

In this example, *all* the differences between people in the characteristic in question—here, chance for being elected—can be summarized by genetic differences between them—that is, possession of either the XX chromosome pair or the XY pair. In this case the heritability of "chance for being elected" would be 1.0. In other words, in this society chance for being elected is 100 percent heritable. But does this by any stretch of the imagination mean that these differences between men and women are genetic? Is there a gene for "electability," a gene that men possess and women do not? Of course, the answer to these questions is no. Although in this case heritability is perfect, it is social (environmental) variables—laws regarding what men and women can and cannot do—that determine whether someone has a chance of being elected. If the law in question were changed and women were allowed to hold office, the heritability of the characteristic of "chance for being elected" would probably quickly fall to much less than 1.0.

The late psychologist Donald Hebb provided another example of this point, drawing on a "modest proposal" put forth by Mark Twain. Hebb notes:

The conception of "heritability" is a misleading one . . . and some of the geneticists who use it are as confused as the social scientists, so its origin in genetics does not guarantee logical use by psychologists. In a 1953 paper, I showed that the amount of variance attributable to heredity (or to environment) cannot show how important heredity (or environment) is in determining an aspect of behavior. . . . I give here a new example. . . .

Mark Twain once proposed that boys should be raised in barrels to the age of 12 and fed through the bung-hole. Suppose we have 100 boys reared this way, with a practically identical environment. Jensen agrees that environment has *some* importance (20% worth?), so we must expect that the boys on emerging from the barrels will have a mean IQ well below 100. However, the variance attributable to the environment is practically zero, so on the "analysis of variance" argument, the environment is not a factor in the low level of IQ, which is nonsense.[54]

Hence, in Hebb's example, environment had no *differential* effect on the boys' IQs; presumably it had the same severely limiting effect in all boys. Because there was no difference—or variation—in the environment, the environment could not be said to contribute anything to differences between the boys. Yet it is also obvious that the environment did have a major influence on the boys' IQ scores. Even with IQ heritability equal to +1.0, the intelligence of each boy would have been different had he developed in an environment other than a pickle barrel. Accordingly, high heritability does not mean developmental fixity. A high estimate of heritability means that environment does not contribute very much to differences among people in their expression of a trait, yet environment may still provide an important (although invariant) source of the expression of that trait, for instance, of the average level of a trait shown by the people in a given group.

In Hebb's statement we see the view that, although heritability may be high, the characteristics in question can be influenced by the environment. Even when environment contributes nothing to the *differences* among people in a population, it does not mean that the population characteristic is fixed by heredity or that it cannot be influenced by the environment. The environment can be a uniformly potent source of behavior development and functioning within each person in a group, while contributing nothing to differences among people.

A point consistent with those above has been made by comparative psychologist Daniel Lehrman.[55] Lehrman noted that when a geneticist speaks of a trait as being heritable, he or she means only that one is able to predict the trait distribution in the offspring of a group on the basis of knowing the trait distribution in the parent group. We can predict the distribution of eye color in the offspring generation merely by knowing the distribution of eye color among the parents. Thus, while the geneticist may use the terms "hereditary" or "inherited" interchangeably with "heritable," it is not a statement about the *process* involved in the development of this trait. Geneticists are not saying anything at all about the way nature and nurture provide a source of a heritable trait, about the extent to which the expression of the trait may change in response to environmental modification. In short, a geneticist would not say that a highly heritable trait cannot be influenced by the environment, but would probably recognize, as we now must, that even if the heritability of a trait is $+1.0$, an almost infinite number of *expressions* (phenotypes) of that trait can develop as a result of interaction with the almost infinite number of environments to which any one genotype may be exposed.

Accordingly, people who equate heritability with genetic determination assume that, as the magnitude of heritability increases from zero to $+1.0$, less and less can be done through environmental modifications to alter how the trait is expressed. Of course, they therefore also assume that if the value of heritability is low there is more room for altering the trait through environmental manipulation. This argument is fallacious. Says Sandra Scarr-Salapatek:

> The most common misunderstanding of the concept "heritability" relates to the myth of fixed intelligence: if h^2 [heritability] is high, this reasoning goes, then intelligence is genetically fixed and un-changeable at the phenotypic level. This misconception ignores the fact that h^2 is a population statistic, bound to a given set of environmental conditions at a given point in time. Neither intelligence nor h^2 estimates are fixed.[56]

So there seems to be some consensus that, despite a high level of heritability, environmental variation may be a (or the) key causative factor. There also seems to be general agreement on a key implication of this view, that high heritability does not mean developmental fixity.

Accordingly, the concept of heritability does not mean what it seems to mean. It says nothing about the genetic basis of a person's behavior; it does not even say much about what it does pertain to. The differences it describes may pertain as much, if not more, to social differences among people as to genetic differences. In any case, if the social context changes, we cannot be certain whether any information we have about heritability still applies.

There are also numerous statistical and methodological problems associated with the determination of heritability.[57] A key problem arises in regard to the concepts of broad heritability (H^2) and narrow heritability (h^2). To understand these concepts, we should recognize that the contributions of variation in heredity and environment to the variance in a given behavior (for example, general intelligence or personality) might be expressed in several ways.

Hereditary and environmental variance can relate to behavioral variance separately and independently; what one contributes is unrelated to what the other contributes. In statistical parlance, such contributions may be labeled "main effects." However, the *contributions* of hereditary variance and environmental variance may "interact." Interaction means that the contributions of hereditary (genotypic) variance to behavioral variance are different under different environmental conditions and that the contributions of environmental variance to behavioral variance vary under different hereditary, or genotypic, conditions.

Still other instances of hereditary and environmental relatedness may be discussed. For example, heredity-environment (or genotype-environment) correlation occurs when hereditary (genotypic) and environmental differences co-vary—that is, when changes in one increase and/or decrease in relation to changes in the other. Together, the concepts of "main," "interactional," and "correlational" contributions of hereditary and environmental variation allow the concepts of broad heritability (H^2) and narrow heritability (h^2) to be discussed and the statistical problems associated with them to be noted.

Broad heritability is the proportion of the total variation in a given behavior that may be attributed to the sum of (1) the independent genetic variation (that is, the main effect of heredity variance) *and* (2) the variance due to genotype-environment correlation. It is of little interest to geneticists, primarily because it does not allow the contribution of genotype variance per se to be disentangled from environmental variance, and of course such separation is a reason to do heritability analyses in the first place.[58]

Narrow heritability(h^2) is simply the proportion of variance in behavior that may be attributed solely to the "main effect" of hereditary (genotypic) variance—that is, h^2 is the hereditary variance that exists independent of, and thus merely separately adds to, environmental variance.[59] Most heritability analysis is aimed at determining h^2, but there lies the rub, especially when analyzing *human* behavior. It is in human heritability analysis and the calculation of h^2 that the statistical and methodological problems of this research arise.

It makes sense to attribute through heritability analysis (and the calculation of h^2) variation in some behavior (for example, in intelligence or in personality) to hereditary variation *only* if the contributions of heredity and of the environment are additive[60]—that is, if heredity and environment interact, then the calculation of heritability becomes quite difficult.[61] When assessing heritability among humans, however, the statistical techniques ("analysis of variance" or its equivalent, the multiple regression model) used to determine whether there is evidence for heredity-environment interaction are not as sensitive to the presence of interactions as they are to the independent (and hence additive) influences of heredity and environment. In other words, these statistical techniques can detect the presence of main effects more readily than the existence of interactions.[62]

This problem is especially apparent when these statistical tests are used with relatively small numbers of observations, and unfortunately such small samples are generally the rule in social and behavioral science studies involving the calculation of heritability.[63] The smaller the sample, the greater the likelihood that the statistical tests will be unable to identify an interaction that is actually present. In such situations, inferences that there is no interaction between heredity and environment and that the contributions of those two factors involve only main (and therefore only additive) effects will be incorrect. As a consequence, heritability estimates and any conclusions based on them are similarly misconceived.

Even if a large sample of observations exists, there are other methodological problems in assessing heritability. For instance, if a scientist wants to determine how much variance in a behavior is accounted for by variance in what people inherit, versus what they experience in their environment, the specific heredity involved in the group under study must be ascertained. If one wants to determine the extent to which variation in children's intelligence is accounted for by the genes provided children by their mothers and fathers, as compared with the environments provided by these parents, one must be certain that one is measuring children and their actual biological mothers and fathers. Thus, as behavior-genetic analysts Jerry Hirsch, Terry McGuire, and Atam Vetta have argued, "A *sine qua non* for the study of heredity is proof positive of the presumed biological relationship, i.e., ascertainment of the biological validity of the designated kinships, such as parent-offspring, sibling, etc."[64]

Unfortunately, few studies of heritability have included such controls for presumed biological relatedness, and therefore one cannot automatically assume that designations of biological relatedness have been accurate in all studies of the contribution of genes to behavioral resemblance between, for instance, children and parents. And even in the few studies that do include such biological controls, there is evidence that substantial proportions of the people labeled as "biological parent" are in fact not related biologically to the children in question. For instance, Hirsch and his colleagues were able to determine, through blood tests, the actual biological relationships between parents and children within a subsample of 38 of the 112 families they were studying. In 13 percent of the families in the subsample, there were children who could not have been the biological offspring of at least one of the putative parents in the family.[65] Similarly, reporting evidence derived from comparable analyses done on samples from the United Kingdom, E. E. Philipp noted that there were data disqualifying as the presumed fathers 30 percent of the husbands within the families being studied.[66]

It is no secret that people mate with people other than their spouses, but when such information remains secret to the researcher who is appraising heritability, literal miscalculations and inferential errors abound. For such reasons, Oscar Kempthorne, an internationally renowned expert in quantitative genetic analysis, has indicated that "most of the literature on heritability in species that cannot be experimentally manipulated, for example, in mating, should be ignored."[67] For these reasons, as well as for the others noted above in regard to heritability analysis, population geneticists Marcus Feldman and R. C. Lewontin have concluded: "Certainly the sample estimate of heritability, either in the broad or narrow sense, but most especially in the broad sense, is nearly equivalent to no information at all for any serious problem of human genetics."[68]

Thus, at best—that is, under very specific conditions—heritability estimates provide limited information. These conditions occur only when there is additive genotypic variation; when there are large sample sizes; when compelling evidence exists for the absence of either a genotype-environment interaction or a genotype-environment correlation; and when there is valid evidence for the biological relatedness of concern in a given anlaysis. Under such conditions, heritability estimates provide only an estimate of the extent to which genetic differences (variation) in a given group are associated with differences (variation) in the scores for a trait measured in people in that group at a specific time in their lives. The presence of such variation in genes and in scores for a trait (e.g., for a given behavior) makes it possible to make two by now obvious points:

1. People in the group in question (for instance, men, women, African-Americans, European Whites, Jews) do not all have the same amount of the trait or characteristic and therefore do not possess an equivalent level of the attribute or characteristic in question. If that were the case, there would be no variation, and heritability would be zero.
2. People in the group do not possess a common genotype. As was the case with the trait, if such commonality existed there again would be no variability and heritability would be zero.

Heritability estimates therefore describe only characteristics of a distribution of scores, only a feature of differences between people. Such estimates tell us nothing about the trait itself. In particular, they say nothing about the genetic and/or environmental determination (or cause) of the trait *within* any person in a group. And certainly, from such an estimate of *between*-people differences (which is, of course, what heritability is) one cannot legitimately make any statements about how humans have been selected for the homogeneous presence of a trait.[69] If such homogeneity existed, heritability would be zero because there would be no variation.

When they use the presence of heritability to support their claims for the genetic determination of universal characteristics of human social functioning, sociobiologists are in effect shooting themselves in the foot. They are in error every time they argue that the presence of heritability supports their beliefs. To the extent that it is useful at all, heritability data actually implies just the opposite. The presence of a heritability estimate greater than zero means that diversity and variation—in genes and in traits—exist within the human group in question.

Psychologist and behavior-genetic analyst Jerry Hirsch has noted that the failure to appreciate this point is a key problem in Wilson's sociobiological conceptions.[70] Hirsch says: "The misleading picture that emerges is that heritability is the very essence of evolution. . . . Wilson [fails] to emphasize the inherent contradiction in this picture, namely, that the important characters [in evolution] have the lowest heritabilities."[71] To underscore his point, Hirsch cites the views of one of the quantitative geneticists Wilson cited in support of his own sociobiological views, D. S. Falconer. Falconer noted: "Characters with the lowest heritabilities are those most closely connected with reproductive fitness, while the characters with the highest heritabilities are those that might be judged on biological grounds to be the least important as determinants of natural fitness."[72]

Given the problems with the heritability concept, any attempt to use heritability data to support a claim for the universal, evolutionarily based genetic determination of behavior is based on a misunderstanding and misapplication of the heritability concept. Hirsch, McGuire, and Vetta summarize the reasons for the problems with using the heritability concept:

> Not only do genotypes differ in response to a common environment, but one genotype varies in response to different environments. Therefore, we need concepts to encompass behavior-genetic relations that are neither isomorphic nor independent; isomorphism might have justified the naive reductionism that led behavior genetics to racism, and independence might have justified behaviorism's naive environmentalism.[73]

The Concept of Adaptability

There is, then, only one line of argument left to support sociobiological claims about the evolutionarily based, genetic source of human social behavior. It involves the concept of adaptation and the view that patterns of human behavior reflect evolutionarily based and therefore naturally selected genetically influenced adaptations to the pressures of the context in which human beings live.

Stephen Jay Gould has noted that a cornerstone of the sociobiological method is to present Rudyard Kipling–like "Just So Stories" about how certain social behaviors, or differences among people in their social status or roles, came to be: "Rudyard Kipling asked how the leopard got its spots, the rhino its wrinkled skin. He called his answers 'just-so stories.' When evolutionists try to explain form and behavior, they also tell just-so stories— and the agent is natural selection. Virtuosity in invention replaces testability as the criterion for acceptance."[74]

According to Gould this unacceptable scientific procedure led the biologist Ludwig von Bertalanffy to complain:

> If selection is taken as an axiomatic and *a priori* principle, it is always possible to imagine auxiliary hypotheses—unproved and by nature unprovable—to make it work in any special case. . . . Some adaptive value . . . can always be construed or imagined. . . . I think

the fact that a theory so vague, so insufficiently verifiable and so far from the criteria otherwise applied in "hard" science, has become a dogma, can only be explained on sociological grounds. Society and science have been so steeped in the ideas of mechanism, utilitarianism, and the economic concept of free competition, that instead of God, Selection was enthroned as ultimate reality.[75]

According to both Gould and von Bertalanffy, then, the key feature of sociobiological "Just So Stories" is that these current arrangements in society are *adaptations*—changes that enhance fitness, that have been shaped by natural selection over the eons of human evolution to have this function, and that are now represented in the human genotype. Yet the key element in the sociobiologist's argument—the presence of an adaptation, of a change in fitness—all too often remains a scientifically unverified post hoc story. Sociobiologist Robin Dunbar frankly admits:

A simple statement that "X increases the fitness of those that perform it" explains nothing: it is strictly tautologous, for improving fitness is what every sociobiological explanation implicitly assumes. What we need to know—and this is the heart of any sociobiological explanation—is: *How* does it increase fitness?

It is the transparent failure to answer this question that has left so many sociobiologists open to criticisms of "Just-So" story-telling and unscientific practice. Since we necessarily have to rely on comparative observations rather than experimental manipulation when tackling evolutionary problems, we are particularly exposed to this kind of accusation. The only way to avoid it is to provide as watertight a case as is possible by showing that proximate problems of survival or reproduction are in fact resolved when individuals behave in a specified way, and that efficient solutions to these problems will result in increased contributions to the species' future gene pool. This will not always be easy, but, unless it can be done, sociobiological explanations will always be open to skeptical doubts, particularly where these doubts are fuelled by political or religious conviction.[76]

Despite these explanatory problems, sociobiologists see adaptations— changes in fitness "designed" by (or, actually, "resulting" from) natural selection—as everywhere. And in the view of sociobiologists, these changes in fitness, because they are adaptations, are *optimizations*. That is, as we have seen in the well-honed arguments of nineteenth-century Social Darwinists and Nazi racial hygienists, natural selection results in genetically based features that are the "time-tested" best possible outcomes of humans' evolutionary history.

To sociobiologists, therefore, that which exists is an adaptation; human social behavior and the niches people occupy in the social hierarchy have been shaped by natural selection to take their present form. As sociobiologist Melvin Konner claims succinctly, "An organism has characteristics; they must have been selected for or they wouldn't be here now."[77] Given the centrality of the concept of adaptation in sociobiologists' thinking, we may question the existence of the direct, uniform, and singular path they infer from evolution through natural selection to adaptation and the present character of people and society. We may ask precisely why presenting a possible scenario of the way natural selection *could have* resulted in a given feature of human behavior is not sufficient to establish scientifically that just such a history transpired.

The work of Stephen Jay Gould and Elisabeth Vrba is quite relevant to these issues.[78] They proposed a new term in evolutionary biology in an effort to clarify some important but confusing uses of the word "adaptation." Gould and Vrba note that one meaning of the term "adaptation" is a feature of the organism (a physical attribute or a behavior, for instance) which is shaped by natural selection for the function it now performs. A second, more static meaning refers to the immediate way in which a physical feature or a behavior enhances the organism's current ability to fit its context. This second meaning does not consider the historical origin of the feature, but only whether the organism's morphological or behavioral characteristics help it meet the current demands present in its environment.

Gould and Vrba cite George Williams as illustrating the first definition of adaptation. Williams contended that one should speak of adaptation only when one can "attribute the origin and perfection of this design to a long period of selection for effectiveness in this particular role."[79] The writings of Walter Bock illustrate the second definition of adaptation.[80] Bock indicates that "an adaptation is . . . a feature of the organism . . . which interacts operationally with some factor of its environment so that the individual survives and reproduces."[81]

Gould and Vrba recognized the confusion over adaptation, a central concept in evolutionary theory.[82] This confusion exists because the same word has been used for two meanings, even though there are different criteria for the historical basis of a given organism feature and for its current use. Darwin himself may have seen the potential for confusion. He said:

> The sutures in the skulls of young mammals have been advanced as a beautiful adaptation for aiding parturition, and no doubt they facilitate, or may be indispensable for this act; but as sutures occur in the skulls of young birds and reptiles, which have only to escape from a broken egg, we may infer that this structure has arisen from the laws of growth, and has been taken advantage of in the parturition of the higher animals.[83]

In other words, while Darwin saw the need for unfused sutures in the skulls of young mammals, he eschewed (as Williams would argue later one should) labeling the unfused sutures as adaptations.[84] As Gould and Vrba point out, Darwin did this because the unfused sutures "were not built by selection to function as they now do in mammals."[85]

But if the unfused sutures are not "adaptations," what are they? Clearly a new term must be used to clear up the confusion. Gould and Vrba came up with such a term. Following Williams, first they define an adaptation as "any feature that promotes fitness and was built by selection for its current role (criterion of *historical genesis*)." They continue:

> The operation of an adaptation is its *function*. . . . We may also follow Williams in labeling the operation of a useful character not built by selection for its current role as an *effect*. . . . But what is the unselected, but useful character itself to be called? Indeed it has no recognized name. . . . Its space on the logical chart is currently blank.

Then they offer their new term, *exaptation*:

> We suggest that such characters, evolved for other usages (or for no function at all), and later "coopted" for their current role, be called *exaptations*. . . . They are fit for their current role, hence *aptus*, but they were not designed for it, and are therefore not *ad aptus*, or pushed towards fitness. They owe their fitness to features present for other reasons, and are therefore *fit (aptus) by reason of (ex)* their form, or *ex aptus*. Mammalian sutures are an exaptation for parturition. . . . The general, static phenomenon of being fit should be called aptation, not adaptation. (The set of aptations existing at

any one time consists of two partially overlapping subsets: the subset of adaptations and the subset of exaptations. This also applies to the more inclusive set of aptations existing through time.)[86]

A clear implication of Gould and Vrba's revised terminology is that not all instances of fitness are adaptations—that is, not all the features of an organism's structure and function that are aptational have this character as a consequence of being shaped by natural selection for this character. If supported, such a possibility would weaken what Gould and Lewontin have labeled the "adaptationist program," the position (reflected in the above-quoted view of Konner) that a feature's current aptational character implies historical shaping by natural selection for that character.[87]

Lewontin has discussed the adaptationist "program" and its conventional use of the concept of adaptation.[88] Like Gould and Vrba, Lewontin sees problems with the "adaptationist program" view of adaptation. He sees the view as deficient because it ignores the active, constructive role the organism plays in its own adaptation: it shapes the context to which it adapts, and hence there is a reciprocal, multilevel (that is, fused) relationship between organism and context. As we shall see in Chapter 6, Lewontin's criticisms of the conventional use of the concept of adaptation derive from a viewpoint that is consonant with the developmental contextual conception of development. Specifically, Lewontin notes that biologists committed to the "adaptationist program" "define adaptation as 'some kind of partial match with the external world' ":

This is the view of a static, or at least independent, outer world that poses fixed problems to which the organism (or population) responds by fitting itself to the preexisting external condition. . . . But life is not like that. Organisms . . . by their own life activities determine which aspects of the outer world make up their environment. Organisms change the environment by their activities. They transduce physical signals from the environment into new physical forms within themselves. Species do not, in general, "solve" preset environmental "problems" by gathering information and responding appropriately. Rather, they "construct" environments. The problem is that the concept of adaptation has been extended metaphorically from its valid domain of describing individual, short-term, goal-directed behavior to other levels. Individual organisms can be observed to be adaptive machines as they steer around obstacles, chase prey, defend themselves from attack, or push their roots around stones. But it is pure metaphor, ideologically molded by the progressivism and optimalism of the nineteenth century, to describe numbers of chromosomes, patterns of fertility, migrations, and religious institutions as "adaptations." It is this kind of error that led to the now discredited descriptions of lemming "migrations" and "mass suicides" as adaptive responses to crowding. It is not simply that some evolutionary process can be described as nonadaptive, but that the entire framework is in question. Whether we look at the fossil record or at living species, we do not see them as "adapting," but as "adapted." But how can that be? How is it that, if evolution is a process of *adapting*, organisms always seem to be *adapted*? It may be more illuminating to see organisms as *changing* and, in the process, as reconstructing the elements of the outer world into a new environment that is sufficient for their survival. To accept the metaphor of adaptation for human culture is to restrict the possible explanations of culture to a relatively narrow range. The price of metaphor is eternal vigilance.[89]

Consistent with the position of Lewontin regarding the problems with the "adaptationist program" view of adaptation, Gould and Vrba contend that if nonaptation "is about to assume an important role in a revised evolutionary theory, then our terminology of form must recognize its cardinal evolutionary significance—co-optability for fitness."[90] In other words, recognition of the potential presence of exaptative, aptative features makes one aware that previously nonaptative (*not* preadaptive[91]) features may be present and may be co-opted for fitness.

Gould and Vrba contend that this co-optation provides a basis for plasticity in evolutionary processes and for the role of an individual's own organismic characteristics in his or her development. That is, they believe there is an "enormous pool" of nonaptations and that must be the source, the "reservoir," of most evolutionary flexibility:

> We need to recognize the central role of "cooptability for fitness" as the primary evolutionary significance of ubiquitous nonaptation in organisms. In this sense, and at its level of the phenotype, this nonaptive pool is an analog of mutation—a source of rare material for further selection.
>
> Both adaptations and nonaptations, while they may have nonrandom approximate causes, can be regarded as randomly produced

with respect to any potential cooptation by further regimes of selection. Simply put: all exaptations *originate* randomly with respect to their effects. Together, these two classes of characters, adaptations and nonaptations, provide an enormous pool of variability, at a level higher than mutations, for cooptation as exaptations [and provide for] . . . the flexibility of phenotypic characters as a primary enhancer of or damper upon future evolutionary change. Flexibility lies in the pool of features available for cooptation (either as adaptations to something else that has ceased to be important in new selective regimes, as adaptations whose original function continues but which may be coopted for an additional role, or as nonaptations always potentially available). The paths of evolution—both the constraints and the opportunities—must be largely set by the size and nature of this pool of potential exaptations. Exaptive possibilities define the "internal" contribution that organisms make to their own evolutionary future.[92]

In essence, the concept of exaptation leads to an understanding of how the processes involved in evolution are plastic ones. Indeed, the concept is consistent with a key theme in the developmental contextual alternative to a biological determinist view of the role of biology in human development: that the person possesses processes that contribute to his or her own plastic functioning, that allow him or her to play a role in the development of his or her own flexibility. Indeed, humans present to their context the exaptative possibilities that provide a key basis for their own future evolution.

THE ABSENCE OF EVIDENCE FOR SOCIOBIOLOGY

The third "line of evidence" of sociobiology—the reliance on an adaptationist "story line" to explain what are purported to be genetically based differences in the social behaviors of women and men—fails. "Just So Stories" about human evolutionary history are used to substitute descriptions for explanations, and alternative paths to current fitness (or aptation) are excluded from scientific consideration or analysis. As was the case with the problems occurring in regard to sociobiology's other two "lines of evidence" (involving the inappropriate postulation of homologies between nonhuman and human animals and misuse of the concept of heritability), the logical and empirical problems with the "adaptationist program" of sociobiology reveal the weak scientific bases of this current instance of the doctrine of biological determinism.

Yet weak scientific justification has not stopped some sociobiologists from making pronouncements about the nature, the genetic basis, of particular social characteristics of specific groups. Just as Nazi-era racial hygienists expressed views about Jews and other "racial" groups that went well beyond what was justifiable on the basis of the scientific data available to them, and at the same time ignored contradictory information and alternative interpretations, sociobiologists of today proceed in a similar fashion. Although current pronouncements do not ordinarily include Jews as a target group, contemporary sociobiologists share with biological determinists of the Nazi era an interest in using their respective ideas to depict the innate basis of the individual and social behavior of two other groups of people: women (and therefore men as well) and Blacks. Some illustrations of the quite comparable depictions of women and Blacks found in Nazi-era social hygienist statements and in some contemporary sociobiological writings will illuminate both biological determinist perspectives. We turn to such comparisons in Chapter 5.

5

BIOLOGICAL DETERMINISM, WOMEN, AND BLACKS: THE PAST AS PROLOGUE TO THE PRESENT

The chief distinction in the intellectual powers of the two sexes is shewn by man's attaining to a higher eminence, in whatever he takes up, than can woman—whether requiring deep thought, reason, or imagination, or merely the use of the senses and hands. [Women have] greater tenderness and less selfishness [but] some, at least, of these faculties are characteristics of the lower races and therefore, of a past and lower state of civilization.
—*Charles Darwin,* The Descent of Man

At least since the time of Charles Darwin, evolutionary theory has contained within it the undercurrents of sexism and/or racism. The Social Darwinist writings of Haeckel and other Monist League members, as well as the writings of German eugenicists, of Nazi-era racial hygienists, and of Konrad Lorenz, were replete with characterizations of other "races" as being inferior and of less value in comparison to the German race. Conflated with this racism was the view that other groups—for instance, criminals, the chronically ill, the physically handicapped, and the mentally retarded—were similarly inferior and possessed lives of little societal or personal worth. Special laws and social policies—ranging from laws that withdrew civil and human rights (such as the Nuremberg laws) through the vast technological, economic, and military coordination required to implement the genocide of the "final solution"—were said by the Nazis and their intellectual forebears and collaborators (e.g., Haeckel, Ploetz, Binding, Hoche, Schallmayer, Lenz, and Lorenz) to be necessary in order to deal with the special problems posed by the presence of such groups in the *volkish* community.

In such ways, then, the science of biologists and physicians working in the historical era surrounding the Third Reich was co-opted by fascist politicians to serve the policy aims of their regime.[1] But that is not a unique historical occurrence—involving only the Nazis or only the doctrines of biological, or genetic, determinism[2]—since scientific ideas may be extended in inappropriate ways to subserve undesirable political aims in any historical period. Moreover, when the science itself is problematic—when the views presented by scholars are themselves flawed in regard to issues of racism, sexism, or "social-classism"—the danger of such political co-optation may be even greater. Despite, or independent of, the personal political views of scholars, their statements in apparent scientific validation of racist, sexist, or classist ideas can provide a ready foundation for political co-option in the service of such ideas.

THE POLITICAL CO-OPTATION OF SOCIOBIOLOGY

The juxtaposition and comparison of the ideas of some contemporary sociobiologists and Nazi-era scientists and politicians in this chapter are not intended to suggest in any way that sociobiologists are neo-Nazis or that they advocate anything akin to the policies promoted within the Third Reich. As already noted, I take sociobiologists at their word when they indicate they do not have such political goals and that they find the attribution of fascist interests to them dismaying and repugnant.[3] However, sociobiologists are aware that critics make the connections between ideas of genetic determinism and Nazi-era ideologies and political activities.[4] Thus, if (as I believe I demonstrate in this chapter) there is congruence between the biological ideas of Nazi-era scientists and contemporary sociobiologists, there is cause for concern when sociobiologists do not speak out against political misuse of their ideas. My point, then, is not to offer an argument for guilt by association, but rather to raise a concern that by ignoring the political context in which their ideas are *used* sociobiologists may *appear* to be validating the use of genetic determinist ideas in racist political agendas.

These concerns are heightened when prominent sociobiologists themselves announce their enthusiasm for the Nazis' racial science, even if not for the Nazis' translation of this science into policy. Such is the case with psychologist and sociobiologist J. Philippe Rushton,[5] discussed later in this chapter. Regarding the scientific and public responses to his opinions, Rushton has asserted: "What the Germans' various final solutions did was deflect racial science from its true course. . . . If it had not been for the Nazi madness these ideas would have continued to evolve ever since 1935. . . . My truer picture of human nature will benefit mankind. I really do believe I have made a major breakthrough in understanding human evolution which can free us from suffering."[6]

There is evidence that other contemporary sociobiologists, such as Wilson, Dawkins, Barash, and Freedman, have messages and methods in common with those of the leaders of the Nazi-era racial hygiene and eugenics movements.[7] Among both the sociobiologists and the Nazi-era scientists, one can find sex-role stereotyped views of the individual and social behaviors of men and women; in both groups, faulty reasoning and inadequate scientific methods are coupled in a manner that is so obviously flawed that the resulting message can hardly be viewed as having only scientific implications. Such presentations by scientists help support a view in society that scholars provide a "truer" picture about genetically determined bases of human social behavior. While the political message was acknowledged by Nazi racial hygienists to be of paramount importance, this is not the stated intent of the above-noted sociobiologists. But as Gould has pointed out, whether a message has implications for politics and/or for social policy is quite independent of the intent or political views of the person presenting the message.[8] Scientists are part of the social context they study, and their pronouncements are also a part of this context.

Sociobiologists Vernon Reynolds, Vincent Falger, and Ian Vine appear to agree with me on this. In the introduction to their *Sociobiology of Ethnocentrism*, they point out that contemporary sociobiological statements have indeed been co-opted by politically right-wing ideologues and that the often inadequate scholarship of their sociobiological colleagues' pronouncements has inadvertently played into the hands of racists who seek to legitimate their ideas by coupling them with "scientific knowledge." Accordingly, Reynolds and his colleagues point out:

> Some attempts to apply the sociobiological paradigm to human behaviour have certainly been premature, ill-considered, superficial, over-confident, and indeed socio-politically naive. Since the political right has already shown its readiness to seize upon mere hypotheses, and to distort them shamelessly before presenting them as "scientific facts" when this can further its case, sociobiologists do perhaps have a special obligation to be cautious in what they say. Because it takes only a relatively subtle misapplication of the general theories to our own species to make it appear that selfishness, and even racism, are "genetic imperatives" of human nature, it becomes essential to stress that nothing of that kind is implied—even if some sociobiologists have seemed to say this.[9]

Moreover, in a related essay, Reynolds both reiterates the *potential* co-optation of sociobiology by the politically far right and identifies the *actual* use of sociobiology for the racist aims of contemporary fascist groups in the United States and Europe: "Scientific ideas are often taken up by politically motivated people, and we have to face the fact that sociobiology has been selectively taken up by the fascist National Front party in Britain and its equivalents in France and the USA."[10]

It is simply not responsible scholarship to make potentially readily inflammable scientific statements (especially statements based on inadequate methodology and/or weak lines of evidence) without considering that these pronouncements are interrelated with the social context in which they are forwarded. Whatever the intent or political persuasion of the scientist, to act as if one's work is somehow independent of the social and political world within which one lives, to act as if science floats in a hermetically sealed world of reason, is at best naive and at worst irresponsible. At the very least, such actions can lead to social mischief, allowing bigotry to claim "neutral" scientific substantiation. At the most extreme, such actions can lead to social policies ranging from discriminatory legislation to the genocide perpetrated by the Nazis.[11]

The potential mischief-through-mayhem continuum associated with the dissemination of weakly reasoned and poorly scientifically grounded statements without regard to the potential for political and social inflammation can be illustrated by reference to issues of sex and racial differences in individual and social behavior. Sociobiologists speak of the genetic basis of behavior differences related to sex and race. These statements of contemporary sociobiologists find a parallel in the views that Nazi racial hygienists expressed about women and about Blacks.

NAZI-ERA AND CONTEMPORARY SOCIOBIOLOGICAL VIEWS OF THE ROLE OF WOMEN

Two groups of people raised quite specific problems for the Nazis. The antirace of the Jews posed a combined genetic, moral, and political threat that impinged on the German *Volk*.[12] Jews were a problem because, if they reproduced, the health of the Nazi state would deteriorate. The other group of people with whom the Nazis had problems was women: if Aryan women failed to reproduce to their utmost biological capacity, the health of the National Socialist state would not be optimized.[13] The need for German women to reproduce as much as possible—to send into the future as many instances of the heroic, masterful Aryan genotype as possible—was seen as so crucial to the survival of the National Socialist state that unmarried women were denied status as citizens of the state. Women were relegated to a subordinate category (*Staatsangehöriger*) to which Jews also were assigned.[14] The assignment of unmarried women to this status was consistent with the view of women that Hitler expressed in *Mein Kampf:* "The German girl is a subject and only becomes a citizen when she marries. But the right of citizenship can also be granted to female German subjects active in economic life."[15] In other words, when full-blooded Aryan women were not in a position to send forth more National Socialists into the next generation, or were not in privileged economic positions, they were relegated to a legal and social status near the bottom of the social hierarchy.

The need for German women to reproduce was so great that Gerhard Wagner, the Führer of the medical profession after the Nazi *Gleichschaltung* ("coordination") of medicine, declared that Germany's "stock of ovaries" was a national resource and the property not of the women "housing" them but of the National Socialist state.[16] Therefore, the Nazis took numerous steps to ensure that the lives of Aryan women would be devoted exclusively to reproduction. For instance, on January 21, 1921, twelve years before Hitler assumed power, the Nazi party at its first general membership meeting adopted a resolution that stated: "A woman can never be admitted into the leadership of the party and into the executive committee."[17] Indeed, in numerous speeches and pronouncements during the Nazi era, Hitler said that the aim of the Nazi party was to "encourage women to marry and stay home."[18] Moreover, Hitler, voicing an opinion markedly akin to the view of Charles Darwin, asserted that a woman's intellect is of "no great consequence."[19]

When the Nazis came into power in January 1933, they began to implement Hitler's plans for women. For instance, in the first few months of National Socialist rule both Jews and married women were dismissed from their teaching positions by administrators of German schools.[20] The reasons for firing the Jewish teachers clearly had to do with the racial and political problems Jews were perceived to pose for German society. The married Aryan women were fired because the Nazis wanted to relegate German women exclusively to domestic tasks.[21] "Even though our weapon is only the soup ladle," declared the *Reichsfrauenführerin* (Reich women's leader),

Gertrud Scholtz-Klink, "its impact should be as great as that of other weapons."[22]

No feature of the Aryan women's domestic role was given greater emphasis in National Socialist Germany than childbearing. The Nazis' emphasis on women as genotype reproducers was so great that they gave awards—the Honor Cross of German Motherhood—to Aryan women who fulfilled their role as producers of new National Socialist citizens—a bronze medal for having four children, a silver medal for six children, and a gold medal for eight children.[23]

At least one leading Nazi-era racial hygienist, Fritz Lenz, believed even eight children was not sufficient to indicate that a woman was doing her duty; such a "low" reproduction record was a sign not of commitment to National Socialism but of the presence of disease and/or political subversiveness (in the Nazi world view, interchangeable problems). Lenz claimed: "It is a fact that a woman is capable of giving birth for a period of nearly thirty years. Even when we consider a woman giving birth only once every two years, this means a minimum of fifteen births per mother. Anything less than this must be considered the result of unnatural or pathological causes."[24]

Given this view of how Aryan women should spend their time, it is not surprising that Lenz believed Aryan women should not waste their time being active citizens in any arena other than reproduction. Not only did he oppose their right to control their own reproduction—to use birth control and/or have an abortion—he also opposed their right to vote, their right to engage in sports, and their right to enroll in higher education.[25] Just as National Socialist physicians were viewed as participating in the Nazi war effort as genetic soldiers, so National Socialist women were also participating in the war as "reproduction soldiers," and, as Hitler asserted, their battlefield was the home.[26]

The way both Lenz and the Nazi Führer saw women is quite comparable to the way contemporary sociobiological perspectives see women. For instance, we have noted Daniel Freedman's claim that because males are selected for aggressive genotypes they competitively garner the world's precious resources of, in particular, ova (or ovaries, as the case may be), and that females are selected for large parental investments—that is, for the reproduction and rearing of children.[27] And according to Richard Dawkins, women have evolved to be exploited by men because of the size of their ova, their gamete carriers.[28] In other words, it is the naturally selected order of the world for women to be objects of exploitation by aggressive, dominant men.

This view of the genetic basis of women's and men's different roles is extended by E. O. Wilson, who links gender differences both to social class and to morality differences, and in so doing posits an association that has a striking parallel to National Socialist racial hygiene ideology. Wilson's 1975 *Sociobiology: The New Synthesis* launched the synthetic discipline of sociobiology with the opening chapter, "The Morality of the Gene," and called for a "genetically accurate and hence completely fair code of ethics."[29] Reflecting both a Social Darwinism as strong as that of Rockefeller or Ploetz and the commission of the naturalistic fallacy, Wilson reveals his belief that genes are the arbiters of fairness and justice: What genes specify in regard to behavior, gender, and social class is fair, since they provide only what natural selection and the struggle for existence have given. The position one occupies in the world is therefore just, since over the course of evolution genes have freely competed for survival. In essence, according to Wilson, the human organs that give us our worth as humans, that allow us to feel and act either good and morally or bad and immorally, "evolved by natural selection," and "that simple biological statement must be pursued to explain ethics and ethical philosophies, if not epistemology and epistemologists, at all depths."[30]

Given this naturalistic ethics of the genes, Wilson sees little possibility that socially instituted freedoms will do anything other than maintain social class differences, especially in regard to vocational roles and particularly regarding individual differences involving men and women: "The genetic bias is intense enough to cause a substantial division of labor even in the most free and most egalitarian of future societies. . . . Even with identical education and equal access to all professions, men are likely to continue to play a disproportionate role in political life, business and science."[31] Thus, Wilson suggests, there may be a genetic predisposition to engage in certain social roles and to enter certain social classes. "Circumstances can easily be conceived by which such genetic differences might occur. The heritability of at least some parameters of intelligence and emotive traits is sufficient to respond to a moderate amount of selection."[32] To illustrate, Wilson notes:

Even in the simplest societies individuals differ greatly. Within a small tribe of !Kung Bushmen can be found individuals who are acknowledged as the "best people"—the leaders and outstanding specialists among the hunters and healers. Even with an emphasis on sharing goods, some are exceptionally able entrepreneurs and unostentatiously acquire a certain amount of wealth. !Kung men, no less than men in advanced industrial societies, generally establish themselves by their mid-thirties or else accept a lesser status for life. There are some who never try to make it, live in run-down huts, and show little pride in themselves or their work. . . . The ability to slip into such roles, shaping one's personality to fit, may itself be adaptive. Human societies are organized by high intelligence, and each member is faced by a mixture of social challenges that taxes all of his ingenuity. . . . The hypothesis to consider, then, is that genes promoting flexibility in social behavior are strongly selected at the individual level.[33]

Wilson provides some specific details about how much genetic action might work: "If a single gene appears that is responsible for success and an upward shift in status, it can be rapidly concentrated in the upper socioeconomic classes."[34]

A view similar to Wilson's is found in the work of Richard Dawkins, whose 1976 book on sociobiology was entitled *The Selfish Gene*. In fact, Dawkins views humans as nothing more than "lumbering robots" containing genes that "created us, body and mind" and states: "Their preservation is the ultimate rationale for our existence."[35] Wilson too believes that this influence on the host-human by the selfish genes means that "the organism does not live for itself." He says: "Its primary function is not even to produce other organisms, it reproduces genes, and it serves as their temporary carrier."[36]

The morality of the gene is therefore not only a selfish, "self-serving" one (if we opt to continue Wilson's and Dawkins's anthropomorphic analogy); the genes also care not whether the behavior they specify the robot to produce is good or bad. As long as the behavior is associated with the reproduction of the gene, the gene is not concerned with whether good or evil, morality or debauchery, is subserved in connection with this replication. Dawkins sees a virtual one-to-one relationship between the selfishness of the gene and the behavior of the person (the "lumbering robot," in his terms) carrying the gene. As a consequence, he claims:

> Like successful Chicago gangsters, our genes have survived, in some cases for millions of years, in a highly competitive world. This entitles us to expect certain qualities in our genes. I shall argue that a

predominant quality to be expected in a successful gene is ruthless selfishness. This gene selfishness will usually give rise to selfishness in individual behaviour.[37]

If the behavior produced by the gene is moral, useful, or healthy, so much the better for society; but the selfish or at best amoral genes could not care less. If the behavior produced is immoral, destructive, or emblematic of disease, so much the worse for society. But again, the selfish gene cares not.

Given such reasoning, questions regarding whether it is fair or humane for people of different social classes or of different genders to be relegated to given social roles are ill-conceived. These differences have evolved and reflect the "natural order" of the world. Such views are interchangeable with the position of Nazi-era racial hygienist Fritz Lenz, regarding the evolutionarily selected, genetically based characteristics of human males and females: "Men are specially selected for the control of nature, for success in war and the chase and in the winning of women; whereas women are specially selected as breeders and rearers of children and as persons who are successful in attracting the male. . . . Hence arise the essential differences between the sexes. . . . Not only do these differences exist, but they are natural and normal."[38]

Dawkins sounds remarkably like Lenz in describing the evolutionarily shaped role of physical attractiveness for women—that is, of the need for women to be physically attractive in order to attract men. Says Dawkins:

It is of course true that some men dress flamboyantly and some women dress drably but, on average, there can be no doubt that in our society the equivalent of the peacock's tail is exhibited by the female, not by the male. Women paint their faces and glue on false eyelashes. Apart from actors and homosexuals, men do not. Women seem to be interested in their own personal appearance and they are encouraged in this by their magazines and journals. Men's magazines are less preoccupied with male sexual attractiveness, and a man who is unusually interested in his own dress and appearance is apt to arouse suspicion, both among men and among women. When a woman is described in conversation, it is quite likely that her sexual attractiveness, or lack of it, will be prominently mentioned. This is true, whether the speaker is a man or a woman. When a man is described, the adjectives used are much more likely to have nothing to do with sex.

Faced with these facts, a biologist would be forced to suspect that he was looking at a society in which females compete for males, rather than vice versa.[39]

The positions of Freedman and Dawkins, on the one hand, and Lenz, on the other hand, are also parallel in that their ideas are based on comparably flimsy scientific evidence. For example, Lenz describes women as dependent and subservient to men, as devoted to their physical attributes and allure, and as socially/interpersonally oriented—without indicating reliance on any research at all (scientifically adequate or otherwise) on personality development in men and women. In turn, Lenz sees men in their by now familiar role as aggressive and heroic agents acting to dominate the "components" of their world, including women:

> A woman wishes, above all, to be regarded as beautiful and desirable, whereas a man wants to be regarded as a hero and as a person who gets things done. Man has more courage than woman in attack, whereas woman shows more valiancy in suffering. Since women are selected by nature mainly for the breeding of children and for the allurement of man, their interests are dependent upon those of man and of children, and are directed towards persons rather than towards things. Owing to the particular nature of her part in life, woman is endowed with more imaginative insight, more empathy, than men. She can more readily put herself in another's place, she lives more for others, her main motive being her love for her husband and her children.[40]

Daniel Freedman's message and method are much the same. In his *Human Sociobiology*, Freedman's style of argument is to present data from one or a few studies and then make generalizations about the human species.[41] Often these studies are found in the appendix, where they are summarized with varying degrees of detail. It is fair to say, however, that there is insufficient information about precisely how the studies were conducted. Crucial elements of scientific methodology are omitted; there is no mention of measurement reliability and validity or of the fact that all the studies in the appendix invariably employed small, unrepresentative samples. In short, the studies Freedman refers to are unpublished, not readily available, and severely underdetailed, and they were not reviewed for scientific quality by Freedman's scientist peers. Yet he combines those studies with his own

opinions and personal anecdotes—and at times with inappropriate inferences drawn from the results of other studies, often of other species—and
makes sweeping generalizations about human beings.

Because these egregious errors in scientific analysis are coupled with
political (e.g., sexist) remarks, a political statement, intentional or not, is
made. The following statements by Freedman about males and females
illustrate my point. Freedman presents them either without empirical documentation or as based on the inadequate studies in his appendix or in some
cases on studies of other species:

> It is by now a well-known fact that women's groups *must* exclude
> men if the average woman participant is to speak openly. The very
> presence of men, however silent they remain, is inhibiting, especially
> to younger women. It can be described as a sort of reflexive "insig-
> nificant little me" response.[42]

> The human male is no exception. It is apparently imperative for the
> male to feel superior to the female—or at least unafraid—for contin-
> uously successful copulations, and it may well be for this reason that
> males everywhere tend to demean women, belittle their accomplish-
> ments, and in the vernacular (clearly laden with symbolism) "put
> them down."[43]

> There appears to be something reflexive in young women that causes
> them to defer to men.[44]

> As far as I can see, the male sense of omnipotence is part of an
> evolutionary heritage among hierarchically arranged species. . . . It
> is the basis for sibling rivalry, for father-son competition, for the
> Oedipus complex, and for the substantial psychological literature
> supporting the existence of that complex.[45]

To get an idea of the kind of evidence on which Freedman bases these
generalizations about the human species, consider the following statement:
"Thus, few fathers achieve the level of empathy with sons that mothers do
(Appendix Study 25) because they are natural competitors."[46] The study to
which Freedman refers (Study 26, *not* Study 25) is entitled "Sex Differences
in Parenting in the Middle Years." It involves the responses of ten young
adults of each sex and their parents to a questionnaire, and Freedman's
summary of that study is quoted here in its entirety:

The study investigated father-child versus mother-child dyadic relationships in the middle years. Ten young adult males and 10 young adult females (mean age 20) completed a questionnaire assessing a range of aspects of the parent-child relationship. Their parents (fathers mean age 47; mothers mean age 45) later were individually interviewed on the same questions as well as on more general adult developmental issues. Measures of the "content" as well as of the degree of "consensus" in parent and child responses were then obtained.

The quality of interaction with children rather than the amount was found to distinguish fathers from mothers. Fathers as compared to mothers appeared less obligated, less varied in active participation in showing care, and less empathetic; they gave and received fewer total indications of affection and showed less mutuality with their young adult children.

Mothers consistently evaluated their interpersonal skills lower than they were rated by their daughters, whereas fathers evaluated their interpersonal skills higher than they were rated by their sons. Such findings of greater discrepancy in same-sex rather than cross-sex parent-child dyads were interpreted, particularly in the case of males, as a manifestation of competition.

The overall results, the author proposed, indicate mothers are more multidimensional and complex in their parenting than are fathers. Mother-child relations are characterized by greater mutuality and reciprocal influencing than are father-child relations. Results, the author showed, are congruent with ego psychological as well as biosocial models of sex differences.[47]

This presentation by Freedman does not qualify as a scientifically adequate report. Freedman provides no details about key features of the study's methodology or about the results of statistical analyses of his "data." Such a report would never be accepted for publication in a scientifically reputable journal. This study, which is representative of the sort of "evidence" Freedman offers in support of his claims, does not provide a scientifically legitimate basis for the formulating of a general statement about the human species. Yet that is precisely what he does.

Presentations like Freedman's represent neither good science nor adequate scholarship. Therefore, when a competent scientist comes up with such conclusions, one wonders whether social and/or political values are involved.

Is such a presentation, cast as science, a responsible one? And are scientists justified in ignoring the potentially inflammatory societal consequences of such statements? The importance of such questions is underscored by the convergence of the views of Nazi-era scientists and some contemporary sociobiologists regarding Blacks and those at risk in society because of the hardships of poverty.

GENETIC DETERMINIST VIEWS OF BLACKS

Although Hitler had a *relatively* better opinion of Blacks than he did of Jews (he is quoted as having said, in response to a question about who should be kept from attending a speech he was to give, "I would rather have one hundred Negroes in the hall than one Jew"[48]), the Nazis' absolute opinion on Blacks and their intellectual forebears was entirely unfavorable. Consistent with what we have seen to be Ernst Haeckel's view—that Blacks were members of an inferior race whose lives were of less value than the lives of members of the higher, German race—Lenz asserted that (using his term) the Negro "certainly lacks foresight"[49] and that this inability is the basis of Negroes' greater tendency to commit crimes, to live in poverty, and to be less apt to work hard to provide for their future.[50] Similarly, Nazi racial hygienist Eugen Fischer stated that the Negro "is not particularly intelligent in the proper sense of the term, and above all . . . is devoid of the power of mental creation [and] is poor in imagination, so that he has not developed any original art and has no elaborate folk sagas or folk myths. He is, however, clever with his hands and is endowed with considerable technical adroitness, so that he can easily be trained in manual crafts."[51]

As with the Nazis' portrayal of women and men, these characterizations of Blacks' genetically based attributes were without any adequate scientific basis. Indeed, today, more than fifty years later, there is still no adequate scientific support for such biological determinist characterizations. Nevertheless, this lack of credible evidence has not stopped some contemporary sociobiologists from characterizing Blacks in virtually the same way as the Nazis did—as slothful, unintelligent, though motorically and physically capable people, who live in crime, poverty, and generally socially deteriorated conditions, and do so because of their genetically based limited mental capacities.

The writings of J. Philippe Rushton are illustrative of these recent sociobiological claims and their comparability with the Nazi conceptions of Blacks.[52] Rushton expresses ideas that are parallel not only to those of Nazi racial hygienists but also to the equally racist writings of the nineteenth-century German biologist Ernst Haeckel.[53]

In an essay on the relationship between principles of sociobiology and race relations, sociobiologist Vernon Reynolds cautions his readers to keep in mind that "the idea that on any absolute scale some races are superior to others is unscientific."[54] He goes on to claim that sociobiology is "explicitly opposed" to such an idea, a conception that should, one would think, be dismissed as laughable because of its thoroughly unscientific status.[55] However, just such an idea—of an absolute scale to judge one "race" as evolutionarily superior to another—has surfaced in the sociobiology literature. And rather than being laughed away, it has repeatedly been able to find its way into print. The idea is the product of Rushton.

Rushton divides the world's population into three "racial" groups, which he terms Caucasoids (Whites), Negroids (Blacks), and Mongoloids (e.g., Asians such as Japanese, Koreans, and Chinese). The key basis of his argument is that "racial differences exist on numerous heritable behaviour traits,"[56] and therefore he relies primarily on one of the problematic "lines of evidence" I have discussed in Chapter 4. Rushton argues that data from several different samples, assessed in several different ways, indicate that for such attributes as brain size and intelligence, maturation rate, personality and temperament, sexual restraint, and quality of social organization one can rank order the three races in similar ways. In essence, Blacks always look the worst—for example, they purportedly have less intelligence and less sexual restraint. In addition, they are portrayed as possessing a more unfavorable social organization—that is, Blacks are described as having more marital instability and greater criminality. Mongoloids (Asians) often come out looking the best, and Whites fall somewhere in between.

Rushton's ideas correspond to those of Haeckel, who contended that the different races within humankind actually represented different species.[57] Indeed, Haeckel characterized Blacks as being comparatively inferior to Whites:

> Now, if instituting comparisons in both directions, we place the lowest and most ape-like men (the Austral Negroes, Bushmen, and Andamans, etc.), on the one hand, together with the most highly developed animals, for instance, with apes, dogs, and elephants, and on the other hand, with the most highly developed men—Aristotle,

Newton, Spinoza, Kant, Lamarck, or Goethe—we can then no longer consider the assertion that the mental life of the higher mammals has gradually developed up to that of man, as in any way exaggerated. If one must draw a sharp boundary between them, it has to be drawn between the most highly developed and civilized man on the one hand, and the rudest savages on the other, and the latter have to be classed with the animals.[58]

Similarly, Rushton says the consistent race differences found across studies arise because Blacks represent a subspecies of humans that is less evolutionarily advanced than Whites and Asians. He contends that Blacks are a group that has a less evolved reproduction strategy—what is termed an r strategy as compared with a K strategy.

An r strategy involves maximization of reproduction (that is, giving birth to large numbers of offspring) with minimization of parental investment (that is, not using much personal energy or emotional, financial, or physical resources for any one offspring). A K strategy involves the reverse: the birthrate is minimized and parental investment is maximized. Rushton argues that Blacks' earlier sexual and physical maturation, their lower levels of sexual restraint, and what he contends are their larger genital size combine with their lower intelligence and their social instability (e.g., their marital discord, poverty, and criminality) to give them a more r-like reproduction strategy than Whites and Asians. The latter racial groups have more K-like strategies, with Asians being more K-like than Whites.

Given the key sociobiological idea—that all people are "lumbering robots" housing "selfish" genes that direct their automaton-like carriers to reproduce themselves[59]—Rushton sees Blacks' more r-like reproduction strategy as accounting for the set of sexual, intellectual, and social-behavioral characteristics he claims are heritable. To be specific:

> Some people are genetically more K than others. . . . The more K a person is, the more likely he or she is expected to come from an intact family, with more intensive parental care, with fewer and more widely spaced offspring, and with a lowered incidence of multiple birthing and infant mortality. K's are expected to have a longer gestation period, a higher birthweight, a more delayed sexual maturation, a lower sex drive, and a longer life.[60]

Ignoring the key fact that r versus K strategies differentiate *species*, not groups (e.g., "races") within species, Rushton proposes—without a shred of direct scientific evidence—that some people, or groups, possess genes that make them more or less K-like. He couples this "science fiction-like" scenario with a list of characteristics purportedly associated with the imaginary "greater K-like genotype." In addition, Rushton ignores the likelihood that most of the variation in these allegedly associated characteristics has social and cultural bases—for instance, pertaining to the economic, political, social service, nutritional, and health care conditions under which generations of Blacks, as compared with Whites, have lived. He then proposes a direct analog between species that vary in r- versus K-like reproductive strategies *and* human groups that purportedly differ in their possession of K strategy genes.

First, Rushton envisions a continuum of reproductive strategies ranging from r to K[61] and locates different species along this continuum—for instance, from "simple" creatures, such as oysters, which produce 500 million eggs a year, to more "complex" animals, such as great apes, which produce only one infant every half decade or so.[62] Given this continuum, Rushton contends that the reproduction strategy Blacks have evolved gives them a less advanced evolutionary status. Thus, in repeating the claim of Haeckel that Blacks are members of a different, subhuman species, the analog Rushton draws depicts Blacks as an evolutionary atavism.

And what is the implication for society of this high rate of reproduction, poverty, family instability, and criminality of Blacks (compared with Whites and Asians)? Within the view of contemporary sociobiologists, and consistent with the opinions of German racial hygienists and eugenicists, such as Ploetz and Schallmayer, the quality of society is diminished when the reproduction of such people is allowed to continue unchecked. For instance, although not specifically mentioning Blacks, statements by Dawkins were aimed at depicting "ignorant" people on "welfare" and appear to have a target that is quite similar to that of Rushton's statements. Writing in his 1976 book *The Selfish Gene*, Dawkins contended that we must be critical of the "unnatural" welfare state, wherein people "too ignorant" to know that they are reproducing more children than they can rear continue to add to the population. Because leaders of the welfare state have removed the safeguards of natural selection, these people are unchecked in their "over-indulgence." Dawkins states:

> Individuals who have too many children are penalized, not because the whole population goes extinct, but simply because fewer of their children survive. Genes for having too many children are just not

passed on to the next generation in large numbers, because few of the children bearing these genes reach adulthood. What has happened in modern civilized man is that family sizes are no longer limited by the finite resources which the individual parents can provide. If a husband and wife have more children than they can feed, the state, which means the rest of the population, simply steps in and keeps the surplus children alive and healthy. There is, in fact, nothing to stop a couple with no material resources at all having and rearing precisely as many children as the woman can physically bear. But the welfare state is a very unnatural thing. In nature, parents who have more children than they can support do not have many grandchildren, and their genes are not passed on to future generations. There is no *need* for altruistic restraint in the birth-rate, because there is no welfare state in nature. Any gene for over-indulgence is promptly punished: the children containing that gene starve. Since we humans do not want to return to the old selfish ways we have abolished the family as a unit of economic self-sufficiency, and substituted the state. But the privilege of guaranteed support for children should not be abused.

Contraception is sometimes attacked as "unnatural." So it is, very unnatural. The trouble is, so is the welfare state. I think that most of us believe the welfare state is highly desirable. But you cannot have an unnatural welfare state, unless you also have unnatural birth-control, otherwise the end result will be misery even greater than that which obtains in nature. The welfare state is perhaps the greatest altruistic system the animal kingdom has ever known. But any altruistic system is inherently unstable, because it is open to abuse by selfish individuals, ready to exploit it. Individual humans who have more children than they are capable of rearing are probably too ignorant in most cases to be accused of conscious malevolent exploitation. Powerful institutions and leaders who deliberately encourage them to do so seem to me less free from suspicion.[63]

Thus, to sociobiologists, the high reproduction rate of the ignorant and the poor represents a threat to society. By "unnaturally" intervening to allow the "surplus" children of the poor to survive, humanitarian programs are actually hurting, not improving, the overall quality of the society. In this regard, and suggestive of the fear of German scientist and Monist League member Wilhelm Schallmayer that society's interference with natural selection, and thus with the struggle for existence, would lead to the destruction of humanity, E. O. Wilson cautions: "If the planned society—the creation of which seems inevitable in the coming century—were to deliberately steer its members past those stresses and conflicts [of the struggle for existence] that once gave the destructive phenotypes their Darwinian edge, the other phenotypes might dwindle with them. In this, the ultimate genetic sense, social control would rob man of his humanity."[64]

Accordingly, the quality of society is at risk if a group within it has evolved to be genetically disposed to have an overly high reproduction rate. This is especially the case if that group purportedly has genetically determined lower intelligence, greater criminality, and less stable and poorer family lives—all characteristics that, Rushton claims, are the case for Blacks. It would seem, then, that in order to support this clearly socially inflammatory argument, it would be the responsibility of a scientist to carefully marshal the highest-quality information. But to defend his line of reasoning, Rushton cites as authorities himself (writing in an earlier theoretical paper that made corresponding claims), Daniel Freedman (writing in his above-described *Human Sociobiology*), and an unnamed French army surgeon writing in his personal journal in 1898.[65]

The quality of the scientific information contained in such sources does not keep Rushton from plunging forward with his claims that the patterns he discerns across the several types of studies reveal unequivocally lower standing of Blacks in regard to all variables he considers. For instance, in discussing the topic of brain size and intelligence, he ignores well-known critiques of the faulty methods involved in using skull size or cranial capacity to estimate intelligence,[66] and he never considers adequately the possibility that group differences in brain size and weight can be influenced by variables (for example, nutrition) tied to socioeconomic status (SES) and cultural factors. Indeed, in discussions of such topics as maturation rate (in regard to group differences in gestation periods or motor development, for example), Rushton never considers the possible confounding role of SES variables.

Similarly, in discussing studies of temperament, Rushton ignores problems of using a psychological device designed to measure the personality of people in one society or culture to assess people of another society or culture. Just as it would be absurd to say that a competent reader of English is illiterate because he or she cannot understand a passage written in Greek, so it cannot simply be assumed that measures of personality devised for people in one culture provide useful information when applied to members of other cultures. Yet Rushton pays no attention to this basic fact of psychological measurement, and he goes on to make pronouncements about, in his terms, the innate temperamental differences among "Negroids," "Caucasoids," and "Mongoloids."[67]

Furthermore, Rushton treats cultural bases of group differences in patterns of sexuality, in family characteristics, and in criminal records as if there were no societal or cultural basis for those behaviors. In his naiveté and/or bias, he writes as if he has discovered something important when he claims that, although Blacks are overrepresented in virtually all categories of crime, they "are underrepresented only among those white-collar offenses that ordinarily require, for their commission, access to high-status occupations (tax fraud, securities violations)."[68] The obvious cultural and societal basis of this "finding" is not mentioned by Rushton, who perhaps is interested merely in marshaling evidence to support the genetic basis of Blacks' behavioral characteristics and of their membership in a human group that he believes is less biologically advanced than Whites or Asians. He therefore can see only additional "evidence" that an evolutionarily advanced reproductive strategy drives the social and sexual behaviors of Blacks.

Rushton also never considers the strength of the findings he reviews—that is, just how much variation between groups is actually accounted for by the differences reflected in the studies he cites. For instance, although in a given study intelligence test scores for Blacks may have a lower *mean* (i.e., arithmetic average) than intelligence scores for Whites—say 85 and 100, respectively—does this demonstrate that there is no variation in scores *within* either group, that, for instance, all Blacks had scores of 85 and all Whites had scores of 100? The answer is certainly no. Variation in scores always exists. Could it be, then, that there was a good deal of variation in both groups—that the scores in the Black group ranged from, say, 70 to 120, while the scores in the White group had a similar range (albeit that the scores for the Whites clustered at a different point within this variation, which is why there were differences in average scores)? The answer to these two questions is definitely yes. Attention to *both* variation and averages makes the task of a scientist more complex, but that is required for an adequate assessment, and a scientist can do no less. By ignoring variation and by focusing only on mean differences, Rushton has not done a complete scientific job. He is telling only part of the story of race differences in physical and behavioral characteristics, and the part of the story he tells is the part that bolsters his claim that the races may be ordered along an r–K continuum.

Similarly, Rushton does not consider how variation in the measures he reviews may be diminished when socioeconomic status is taken into account. That is, to what extent may the differences between Blacks and Whites be based on socioeconomic status rather than on race, and if one could equate the groups on SES would the differences disappear? For instance, if the results of a test of spelling ability showed that fifteen-year-old girls spelled better than five-year-old boys, one would *not* conclude that girls were better spellers than boys; the differences between the groups could be attributed to age differences as readily as to sex differences. In this example of a poor and inadequate study, the different effects "sex" and "age" might have on spelling cannot be separated and therefore are said to be confounded—a good study would use boys and girls of the same age. Similarly, a good analysis of race differences would always take into account that Blacks are more likely to be in lower socioeconomic groups than Whites. There are ways to control for (or "remove") the influence of socioeconomic status on the measures Rushton discusses, but such procedures do not find their way into his work, which remains focused on analyses that are too simplistic and confounded. Once again, Rushton is content to concern himself only with discussing mean differences between groups and to ignore any variability surrounding a mean. Thus, he never discusses the degree of absence or presence of overlap in the distributions of scores for a given characteristic. Simply, to Rushton, if a mean difference exists, it indicates without exception that one group—as a whole—is different from the other. Such naive and scientifically improper use of what are quite basic statistical methods, coupled with the misuse of heritability data, suggests not only poor scholarship but also an attempt to force a political as well as a scientific point.

Yet Rushton sees his work in a quite different light. I have already quoted him as characterizing his scholarship as constituting a "major breakthrough," as providing a "truer picture of human nature [that] will benefit mankind."[69] But what is this "breakthrough," this "truer picture"? It is ultimately the same picture envisioned by Nazi racial hygienists and ideologues—that racial disease can be cured by recognizing the degenerating effect people with inferior, mongrelizing genes have on the Best of the nation. And how can the Best be freed from the suffering that Rushton indicates exists? Only by enacting social policies that keep people with inferior genes "in their place" and eliminate the possibility that the behaviors of the superior will be "infected" with the less evolutionarily advanced, genetically based characteristics.

The cure Rushton envisions may free some people from suffering, but if the Nazi era taught us anything, such a cure will only come by inflicting pain on others, a pain which—because of the pervasive character of social life which will have to be organized in order to enforce the "cure" successfully—will eventually engulf and destroy both those who receive it and those who inflict it. Rushton's thinking, so redolent of Nazi-era political and scientific pronouncements about advances in cures of genetic disease, is nothing more than the most recent instance of genetic determinist ideology promoted as science. His work, and that of many other contemporary sociobiologists, is poor science and represents a fatally flawed basis for prescribing social policy. Scientists and citizens alike must confront both these domains of shortcomings. If we do otherwise, we are allowing history to repeat itself.

TOWARD A NEW ROLE FOR BIOLOGY

What are we to make of the message of sociobiology, of this newest instance of biological determinist ideology? Given the consistency in the "scientific" agenda of contemporary sociobiologists and that of earlier German racial hygienists, and given (at least in regard to Rushton) the homage paid to the racial science ideas developed in National Socialist Germany, it may be difficult for people to see the work of contemporary sociobiologists as apolitical and as not intended for social policy use but only for dispassionate scholarly debate, as is at times claimed.[70]

For instance, E. O. Wilson notes that he and other scholars who have "a similar belief in the biological conservatism of human nature" "can scarcely be accused of having linked arms to preserve the status quo, and yet that would seem to follow from [what he regards as] the strange logic employed" by his critics (e.g., the Sociobiology Study Group of Science for the People). He concludes:

All political proposals, radical and otherwise, should be seriously received and debated. But whatever direction we choose to take in the future, social progress can only be enhanced, not impeded, by the deeper investigation of the genetic constraints of human nature, which will steadily replace rumor and folklore with testable knowledge. Nothing is to be gained by a dogmatic denial of the existence of the constraints or attempts to discourage public discussion of them. Knowledge humanely acquired and widely shared, related to human needs but kept free of political censorship, is the real science for the people.[71]

Although there should indeed be open debate about scientific issues, and the ethical acquisition and dissemination of knowledge should be free of political censorship, such freedom includes the right to point out that, or question whether, purportedly scientific statements actually have a political component. Serious violations of the rules of scientific debate—parading speculation, opinion, or bias as fact; using findings from a few studies to make generalizations about the entire human species; making uncritical comparisons among studies of very different species; presenting descriptive similarity uncritically as well-documented homology; offering description as explanation; misusing and misinterpreting basic principles of statistics and measurement; combining anecdote with information from unpublished, inadequate studies to document sociobiological ideas—are abuses that are not compatible with the very style of scientific investigation and open debate for which Wilson appeals. When so many of the rules are so blatantly violated, one cannot help but wonder whether the message is actually aimed at a political, not a scientific, audience.

But sociobiologists, as writers claiming to be conducting the business of science, often criticize their critics for being merely destructive, for attempting to refute sociobiology without offering a viable alternative.[72] They also claim that the objections are based merely on a knee-jerk reaction to any theory that stresses the role of biology in human behavior. However, neither of these characterizations of the critics of sociobiology is appropriate.

It *is* possible both to include in one's message about human behavioral development a conception of the ubiquitous role of biology and not resort to genetic reductionism, biological determinism, and political and social bigotry. One can offer another version of how biology plays a role in human behavior and still not resort to the "straw men" interpretations Wilson poses as "alternatives" to sociobiology.[73] A conception of human behavioral development can offer (1) a scientific alternative to biological determinism which at the same time emphasizes the ubiquity of the role of biology in all human behavior and (2) a quite dramatically distinct set of social policy implications. As noted in earlier chapters, this type of conception sees human biological and psychological *development* in a synthetic relationship with features of our social world, the institutions of our *social context*. Consistent with Wilson's call for open discussion,[74] I turn now to a presentation of this developmental contextual view of human life.

6

DEVELOPMENTAL CONTEXTUALISM: BIOLOGY AND CONTEXT AS THE LIBERATORS OF HUMAN POTENTIAL

[Our] conception of human development . . . differs from most Western contemporary thought on the subject. The view . . . is that humans have a capacity for change across the entire life span. It questions the traditional idea that the experiences of the early years, which have a demonstrated contemporaneous effect, necessarily constrain the characteristics of adolescence and adulthood. . . . There are important growth changes across the life span from birth to death, many individuals retain a great capacity for change, and the consequences of the events of early childhood are continually transformed by later experiences, making the course of human development more open than many have believed.
—Orville G. Brim, Jr., and Jerome Kagan
Constancy and Change in Human Development *(1980)*

Biological determinism is a pessimistic and potentially socially pernicious doctrine. It is pessimistic about the possibility that the physical or social environment can alter the quality of human life, solve social problems, or free people from the constraints on their behavior. It sees biology as a factor that puts limits on human performance and potential. In this view, biology represents the natural world's inevitable and immutable constraint on the individual and on the organization of society.

Throughout this book I have illustrated the pernicious implications this conception of biology can have for social policy and for the lives of millions of human beings. But because a doctrine is pessimistic does not mean it is wrong. Because a doctrine has undesirable social-policy implications does not mean it is scientifically invalid. However, when a doctrine is scientifically flawed, the pessimism it conveys about the human condition may be unwarranted, and the negative policy implications derived from it may not be justified.

I believe these problems exist with the doctrine of biological determinism in general, and in particular with genetic determinist instances of this doctrine such as represented by sociobiology. I believe that the conception of biological processes, and indeed of human developmental processes in general, represented in sociobiological thinking—and in the earlier, Haeckelian, German racial hygiene and eugenic, and Lorenzian ethological positions—is scientifically flawed.

HEREDITY AND ENVIRONMENT FUSIONS
IN HUMAN DEVELOPMENT

Biological determinism is not the only conception of biological or genetic functioning. A considerably more optimistic view, developmental contextualism, exists. Here the ever-present influences of biology on human behavior and development are acknowledged, as in biological determinism doctrines, but there is a key difference in how the contributions of biology are envisioned.

In the developmental contextual view, genes do not act independently of the environment; they are neither the primary nor the ultimate causal influence on behavior. Instead, heredity and environment are seen as co-equal forces in the determination of behavior. Indeed, those two domains—heredity (genes) and environment—are seen to be completely *fused* in life. Thus, developmental contextualism sees biological and contextual *levels* as integrated. Each level, in developmental contextualism, is an organization of a specific set of phenomena. For instance, biology is one level of organization and involves such entities or phenomena as genes, hormones, and nerve cells. The individual is another level and is comprised of such characteristics as motivation, personality, and intelligence. Society and culture are still other levels and involve such phenomena as families, communities, educational systems, and industrial and technological institutions.

Within developmental contextualism the "conditions" of a specific integration—that is, the particular levels fused in respect to a given characteristic of the person—provide the basis for behavior or development.[1] Genes always exist in an environment, and the specific characteristics of this context influence their functions. For example, under most environmental conditions human genes are involved in the development of two arms and two legs, but unfortunate cases in the 1950s, when scores of women took the drug thalidomide during their pregnancy, showed that even a slight change in the chemicals in an unborn baby's prenatal environment can result in an infant born without any limbs at all.

The environment would be an empty and meaningless place if there were no people in it, and the influence the environment has on behavior depends, then, on the specific characteristics—the specific hereditary makeup—of the people who live in it. For instance, no known type of environmental enrichment (medical or nutritional) can halt the emergence of Tay-Sachs disease (infantile amaurotic idiocy) in infants who have the single recessive gene for that disorder, and as far as we know, no environmental agent (for example, a virus) can produce the disorder. However, sickle-cell anemia, which like Tay-Sachs disease is a single-gene, recessive condition, can in some instances be treated through bone marrow transplants from a suitable sibling donor.[2] Rapid advances in recombinant DNA technology and thus in genetic engineering make it possible that in the not-too-distant future we may be able to use genes to manufacture, in the blood of children with sickle-cell anemia, biochemicals that will treat the disease.[3] Thus examples of the current and future potential health implications of single-gene recessive disorders illustrate the point that the specific characteristics of the integration of events and processes from external (for example, societal) contextual levels (including those associated with scientific discovery and medical technology) and from internal, biological processes of the person shape present and possible future developmental trajectories.[4]

The fusions of heredity and environment—of nature and nurture—mean that they are mutually permissive *and* mutually constraining in influencing behavior. Biology may "permit" more or less of a given behavior and/or may promote one or another quite different behavior, *depending on the environmental circumstances within which people exist and the levels of person and context whose integration is involved in a particular characteristic.* For example, a girl may have genes that are associated with beginning the menstrual cycle quite early, say at about ten years of age, but the nutrition and health care she receives will influence whether her cycle begins then, later, or perhaps even earlier. In turn, the environment may "promote" more or less of a behavior, and/or may bestow one or another characteristic, *depending on the specific biological characteristics of the people living in the environment and, again, the levels of person and context whose integration is involved in the characteristic in question.* For instance, excellent nutrition and health care may maximize the possible height of people who are of hereditarily shorter stature than the average person (for example, members of pygmy tribes), but no known diet or medical intervention will raise the typically occurring heights of members of this group to magnitudes found, say, in groups having hereditarily tall stature.

Thus, genes and environment always constrain each other, but their mutual influence on each other means that these constraints are flexible, not absolute. The human genotype (or genome) constrains people's ability to see through the skull of another person in order to inspect the brain for lesions or tumors; however, by participating in the development of the cognitive system of human beings, their ingenuity and industriousness, the same genome contributes to people's ability to peer into the brains of others through the invention and production of X-ray and CAT-scan machines.

THE CONCEPT OF RELATIVE PLASTICITY

The developmental contextual perspective is consistent with geneticist R. C. Lewontin's views about the issue of constraints, published in a 1981 article:

> It is trivially true that material conditions of one level constrain organization at higher levels *in principle*. But that is not the same as saying that such constraints are quantitatively nontrivial. Although every object in the universe has a gravitational interaction with every other object, no matter how distant, I do not, in fact, need to adjust my body's motion to the movement of individuals in the next room. The question is not whether the nature of the human genotype is relevant to social organization, but whether the former constrains the latter in a nontrivial way, or whether the two levels are *effectively* decoupled. It is the claim of vulgar sociobiology that some kinds of human social organization are either impossible, or that they can be maintained only at the expense of constant psychic and political stress, which will inevitably lead to undesirable side effects because the nature of the human genome dictates a "natural" social organization. Appeals to abstract dependencies (in principle) of one level or another do not speak to the concrete issue of whether society is genetically constrained in an important way. . . . In fact, constraints at one level may be destroyed by higher level activity. No humans can fly by flapping their arms because of anatomical and physiological constraints that reflect the human genome. But humans do fly, by using machines that are the product of social organization and that could not exist without very complex social interaction and evolution. As another example, the memory capacity of a single individual is limited, but social organization, through written records and the complex institutions associated with them, makes all knowledge recoverable for each individual. Far from being constrained by lower-level limitations, culture transcends them and feeds back to lower levels to relieve the constraints. Social organization, and human

culture in particular, are best understood as negating constraints rather than being limited by them.[5]

In short, then, the fusion of heredity and environment as conceived within developmental contextualism, and the resulting mutuality of influence and flexibility of constraints which derive from this fusion, mean that there is *relative* plasticity in human behavior and development: the range of behaviors that can occur in an individual's life is certainly not infinite or limitless. Females cannot, as a group, be changed to begin their menstrual cycle (that is, to experience menarche) at five years of age, and pygmies, as a group, cannot be changed to have an average adult height of six feet. However, the concept of relative plasticity means that the number of distinct behaviors any one individual can show is quite large given the *fusion* of heredity and environment—that is, the particular levels whose integration are involved in a given characteristic and the relative constraints (or degrees of plasticity) involved in these levels. For example, girls' menarche can occur, within normal limits, anywhere between the ninth and seventeenth years,[6] the average adult height of any group can vary widely, and the intelligence, personality, or motivation of people can show an enormous degree of variation.

Given the character of relative plasticity, then, there is extensive variability among people—because genotypes and environments are almost infinitely varied and no two people in the world have the same fusion of genes and environments across their lives.

Even identical (monozygotic) twins—that is, twins born from one fertilized egg that split into two separate fertilized eggs (zygotes) after conception—do not share the same environments across life. Environments differ in their physical, community, social, and cultural characteristics, and although identical twins have the same genotypes, their respective genes are not likely to be fused with identical environments across their entire life spans. Each twin meets different people, may have different teachers, and will fall in love and marry a different person. As such, even for identical twins, behavior and development will be different.

Simply, there are multiple levels of the environment, or context, of life, and differences within each level. For example, within the physical environment there are differences in noise, pollution, climate, and terrain. Genotypes are at least equally variable. The fusion of these two sources of human behavior and development—the environment and the genes—means that, in effect, *each person is distinct* from all other people. The magnitude of individual differences among people underscores the gross errors one makes when characterizing entire groups of people—racial, religious, or gender groups—as homogeneous and undifferentiated in significant ways. It is important to consider how our biology in fact *ensures* systematic individual differences among people.

GENETIC BASES OF INDIVIDUAL DIFFERENCES

Our genetic endowment provides a basis for the uniqueness of each human life and gives substance to the claim that all humans have a unique history of heredity-environment fusions.[7] Estimates of the number of gene pairs in humans typically range between 10,000 and 100,000.[8] Indeed, as pointed out by quantitative geneticist Gerald McClearn, if one considers how much genotype variability can be produced by the reshuffling process of meiosis (the division that forms our sex cells—our sperm and our ova) occurring with 100,000 gene pairs, the potential for variability is so enormous that "it is next to impossible that there have ever been two individuals with the same combination of genes."[9]

Indeed, a conservative estimate is that there are more than 7×10^{17} (or over 70 trillion) potential human genotypes.[10] And geneticists have estimated that each human being has the capacity to generate any one of $10^{3,000}$ different eggs or sperm; by comparison, their estimate of the number of sperm of all men who have ever lived is only 10^{24}.[11] Thus, according to McClearn, considering "$10^{3,000}$ possible eggs being generated by an individual woman and $10^{3,000}$ possible sperm being generated by an individual man, the likelihood of anyone—in the past, present, or future—ever having the same genotype as anyone else (excepting multiple identical births, of course) becomes dismissibly small."[12]

If we recognize that *genetic* does not mean *congenital*, a given human's genetic individuality can be seen as even greater. The "total genome is not functioning at fertilization, or at birth, or at any other time of life."[13] Therefore, the expression of any individual human genotype is a developmental phenomenon, influenced in regard to the turning on and/or off of genes by the internal and external components of the individual's history of genotype-environment fusions.[14] McClearn uses "the differential production of certain kinds of hemoglobin during various phases of development" as an illustration, saying, "for example, production of the beta chain accelerates at the time of birth and peaks after a few months, whereas production of the alpha chain rises prenatally and maintains a high level."[15] Still another indication of the possible variability among humans is the nature of the molecular structure of genes. An estimated 6 billion nucleotide bases comprise the DNA material of the human genome.[16] The vast number of these distinct chemicals constitute an enormous "population" within which mutation—permanent alterations in genetic material—can occur.

This enormous genetic variability among humans is all the more striking because in the determination of behavior it is fused with environments, which have at least equal variation. Thus, fusion means that heredity and environment do not function separately in the determination of behavior and development. Furthermore, they do not merely interact, because "interaction" connotes two independent entities that merely multiply in their effects on behavior.[17] Rather, fusion implies a reciprocal relation between components of an intermeshed system, interactions that are termed *dynamic*.[18] As comparative psychologist Gilbert Gottlieb explained in 1991, the most significant feature of this dynamic, systems view "is the explicit recognition that the genes are an integral part of the system and that their activity (i.e., genetic expression) is affected by events at other levels of the system, including the environment of the organism."[19]

Accordingly, as comparative psychologist Ethel Tobach has emphasized for more than two decades, genes *must* dynamically interact—be integrated or fused—with the environment if they are to be involved in the development of *any* physical or behavioral characteristic of a person.[20] In this vein, Gottlieb explains:

> Genetic activity does not by itself produce finished traits such as blue eyes, arms, legs, or neurons. The problem of anatomical and physiological differentiation remains unsolved, but it is unanimously recognized as requiring influences above the strictly cellular level (i.e., cell-to-cell interactions, positional influences, and so forth). . . . Thus, the concept of the genetic determination of traits is truly outmoded. . . . Although we do not know what actually causes cells to differentiate appropriately according to their surround, we do know that it is the cell's interaction with its surround, including other cells in that same area, that causes the cell to differentiate appropriately. The actual role of genes (DNA) is not to produce an arm, a leg, or fingers, but to produce protein (through the coactions inherent in the formula DNA $\leftrightarrow$ RNA $\leftrightarrow$ protein . . .). The specific proteins

produced by the DNA-RNA-cytoplasm coaction are influenced by coactions above the level of the DNA-RNA coaction. . . . [For example,] an extraorganismic environmental event such as a brief period of exposure to ether occurring at a particular time in embryonic development can alter the cytoplasm of the cell in such a way that the protein produced by the DNA-RNA-cytoplasm coaction eventually becomes a second set of wings . . . on the body of an otherwise normal fruitfly. Obviously, it is very likely that "signals" have been altered at various levels of the developmental hierarchy to achieve such an outcome.[21]

Given, then, that the influence of genes depends so thoroughly on *when*, in the life of the organism, they coact with the environment and with *what* environmental experiences genes coact at that time, it is important to understand how the dynamic integration of genes (heredity) and environment provides a basis for the *relative plasticity* of behavior.

THE DYNAMIC INTERACTION OF HEREDITY AND ENVIRONMENT

A key basis of the doctrine of genetic determinism is the idea that, in trying to understand the basis of a person's behavior, one may separate genes from environment. In other words, genetic determinists are in effect asking which—nature or nurture, heredity or environment—is the cause of an individual's behavior and development. Their answer is inevitably "heredity." However, to ask the "which" question is to muster the wrong army for the battle at hand. As indicated earlier, without heredity there would be no one *in* an environment, but no place to see the "effects of heredity" without environment. Indeed, in the developmental contextual perspective it is even inappropriate to speak of the "effects of heredity" because genes and contexts are inextricably fused; it is actually the *characteristics* of this integration of biological and contextual levels that provide behavioral outcomes or "effects" in development. In other words, to speak of the "effects of heredity" is incorrect in this view because it connotes that heredity influences are somehow separate from the contextual levels with which they are integrated.[22] Genes do not exist in a vacuum. They exert their influence on behavior in an environment that has specific characteristics at each of the several levels: physiological, social, cultural, and so on. At the same time, if there were no genes (and consequently no heredity) the environment would not have an organism in it to influence.

Accordingly, nature and nurture are inextricably tied together. In life they never exist independent of the other. Therefore, in order to be logical and to reflect life situations accurately (that is, to have *ecological validity*), *any* theory of development must stress that nature and nurture are always involved in all behavior. It is simply not appropriate to ask which, because they are both necessary for any person's existence or for the existence of any behavior.

Instead, one should ask *how*: "How are nature and nurture integrated to provide a basis for behavior?"[23] The multilevel processes of the dynamic integration between heredity and environment provide an answer to this question.[24] To illustrate how these processes function, I represent hereditary contributions by the letter G (for genes), environmental contributions by the letter E, and behavioral outcomes by the letter B. It is possible to think of heredity's contribution to behavior as being "direct," as not being influenced by the environment (see Figure 2). That is the idea found in biological determinism. In this formulation, labeled (a) in Figure 2, a particular combination of genes (G_1) will invariably lead to a particular behavioral outcome (B_1). However, I have argued that this view is not congruent with the facts of the real world. The alternative, developmental contextual view is illustrated in Figure 2 as (b). Here the same hereditary contribution (G_1) can be associated with an infinity of behavioral outcomes $(B_1$ to $B_n)$ as a consequence of integration with the infinity of environments $(E_1$ to $E_n)$ that could exist.

Numerous examples exist of the dynamic integration of biological (e.g., genetic) and environmental levels illustrated by Figure 2(b). For instance, under experimentally altered rearing conditions, it is possible to produce mammal-like teeth in birds, which under their typical conditions of development never have teeth.[25] Another, quite striking example of gene-environment dynamic interaction occurs under what *are* typical conditions of development for coral reef fish. Gilbert Gottlieb says:

These fish live in spatially well-defined, social groups in which there are many females and few males. When a male dies or is otherwise removed from the group, one of the females initiates a sex reversal over a period of about two days in which she develops the coloration,

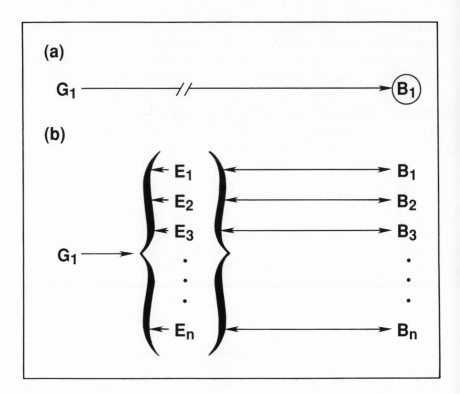

Figure 2.   (a) Genes (G) do not directly lead to behavior (B). (b) Rather, the influence of the genetic level will vary when integrated with (1) levels of the environment (E) having different characteristics; and (2) the behavioral level.

behavior, and gonadal physiology and anatomy of a fully functioning male. . . . Such sex reversals keep the sex ratios about the same in social groups of coral reef fish. Apparently, it is the higher ranking females that are the first to change their sex and that inhibits sex reversal in lower ranking females in the group.[26]

This example illustrates how the expression of genes at hormonal, physiological, and morphological levels changes as a function of the presence of different characteristics within the changing abiotic and biotic levels of the context with which they are integrated. In particular, it appears that the changing number of fish in the group affords the different sources of stimulation involved in the occurrence of the changes in sex.[27]

There are also examples of the integration of genetic and environmental levels among humans. Consider the case of a child born with Down's syndrome. The genetic material—the DNA—of genes is arranged on string-like structures present in the nucleus of each cell. These structures are chromosomes. The typical cells of the human body have forty-six chromosomes, divided into twenty-three pairs. The only cells in the body that do not have forty-six chromosomes are the gametes, the sex cells (sperm in males, ova in females). These cells carry only twenty-three chromosomes, one of each pair. This arrangement ensures that when a sperm fertilizes an ovum to form a zygote, the new human so created will have the number of chromosome pairs appropriate for the species. However, in a child born with Down's syndrome, a genetic anomaly exists. There is an extra chromosome in the twenty-first pair: three chromosomes are present instead of two.

Thus, children with Down's syndrome have a specific genetic inheritance. The complement of genes transmitted to people at conception by the union of the sperm and the ovum is termed the *genotype*. This is what constitutes our genetic inheritance. At least insofar as the extra chromosome is concerned, the Down's syndrome child has a specific genotype. Yet even though the genotype remains the same for any such child, the behavioral outcomes associated with this genotype differ.

As recently as thirty years ago, Down's syndrome children, who are typically recognized by certain morphological (particularly facial) characteristics, were expected to have life spans of no more than about twelve years. They were also expected to have quite low IQ scores. They were typically classified as a group of people who, because of their low intelligence, required custodial (usually institutional) care. Today, however, Down's syndrome children often live well beyond adolescence and also lead more self-reliant lives. Their IQs are now typically higher, often falling in the range allowing for education, training, and sometimes even employment. Indeed, there is at least one quite striking instance of substantial intellectual accomplishment by a person with Down's syndrome. Nigel Hunt, born with Down's syndrome, wrote his autobiography.[28]

Of course, plasticity is relative, not absolute, and as such there are limits in the expectations we might appropriately maintain about the life-span development of a person with particular characteristics of organismic individuality, such as Down's syndrome. Nevertheless, across even recent history there are major differences in the functional implications of such genetically related disorders,[29] and it is therefore appropriate to ask how these vast differences in the behavioral impact of the Down's syndrome genotype came about. Certainly the genotype itself did not change. What did change was the environment of these children. Instead of invariably being put into institutions, different and more advanced special education techniques were provided, often on an outpatient basis. These contrasts in environment led to variation in behavioral outcomes despite the same heredity—that is, despite *genotypic invariance*.

There are also illustrations of the fact that heredity always exerts its effects "indirectly"—through its fusion with the environment, in the development of physical as well as behavioral characteristics. First, consider the disease phenylketonuria (PKU), which involves an inability to metabolize certain fatty substances because a particular digestive enzyme is missing. This disease leads to the development of distorted physical features and severe mental retardation in children. Because the missing enzyme resulted from the absence of a particular gene, PKU is another instance of a disease associated with a specific genotype. Today many people—perhaps even some reading this book—may have the PKU genotype without having either the physical or the behavioral deficits formerly associated with the disease. It was discovered that if the missing enzyme is added to the diets of newborns identified as having the disease, *all* negative effects can be avoided. So again, changing the environment changed the outcome. Therefore, we can see that the same genotype will lead to different outcomes, both physical and behavioral, when integrated with contrasting environmental settings.

More striking than these examples of dynamic integrations of genes and environment are those indicating that *environment can actually alter key features of the genotype itself: environment can alter the function of the DNA*. The results of several experiments indicate that exposure to complex and stimulating environments—that is, enriched contexts, as opposed to simple, unstimulating, or impoverished settings—alters the functions of the most basic chemical constituents of genes, the carriers of the genetic message itself: the RNA (ribonucleic acid) and the DNA (deoxyribonucleic acid). In one study rats were exposed to high environmental enrichment (living in a cage with eleven other rats and having "toys" and mazes available for exploration), low environmental enrichment (living in a cage with one other rat but no exploration materials), or isolation (living in a cage alone and with no exploration materials). When the brains and the livers of the rats were studied to determine the transcription to DNA by RNA, the RNA from the brains of the environmentally enriched rats showed a level of transcription of DNA significantly greater than that of the other groups; no significant differences were found with liver RNA.[30] Similarly, other investigators have found significant differences between the brain RNA of rats reared in environmentally rich versus environmentally impoverished contexts.[31] In addition, the total complexity of brain RNA has been found to be greater for normally sighted kittens than for kittens who had both eyelids sutured at birth.[32] However, because the RNAs from the nonvisual portions of the cerebral cortex and from subcortical brain areas were not different for the two groups, the normal development of the visual cortex in this species, which is dependent on visual experience, appears to involve a greater amount of genetic expression than occurs in the absence of visual experience.

We should remain cautious about extrapolating specific findings about nonhuman processes to processes involved in humans. Nevertheless, it seems clear that the dynamic integration between genetic and environmental levels can alter both the behaviors that develop across life and the chemical messages involved in the genes themselves. Indeed, the data in support of the integrations represented in Figure 2 weaken (to the point of complete untenability) any claim that genes represent a blueprint for a single outcome of development. In other words, the data in support of the figure show that there is no ismorphism between genotype and phenotype. Thus, any such "extreme" genetic determinist claims could not be supported in the face of this evidence. Similarly finding no support in relation to such evidence would be another, albeit "weaker," but still genetic-determinist version of the claim that genes directly translate into behavior.

For instance, it might be argued that genes are associated with a limited range of outcomes rather than a single behavioral outcome, and also that for people from genetically different groups (for example, women and Blacks) this range would not overlap (either at all or even appreciably), particularly given the "average expected environments" these different people might encounter. Such an argument might be predicated on the fact that the genetics literature often stresses that for a given genotype there is an array of outcomes, of phenotypes, that can be associated with the genotype, depending on the specific environmental characteristics in which the genotype develops.[33] This array of outcomes is termed the "norm of reaction" in the genetics literature.[34] The presence of such a norm is consistent with the fusions represented in Figure 2; both the figure and the norm of reaction concept underscore the plasticity that exists as a consequence of the integration of multiple levels of organization, including the genetic level. However, it is precisely because of this plasticity that the concept of norm of reaction cannot be used to support even the above-noted weaker genetic determinist position. As behavior-genetic analyst Jerry Hirsch emphasizes, for any living form the norm of reaction remains largely unknown,[35] because to be able to specify exactly the norm of reaction for any living animal (or plant) one must be able to reproduce exactly an individual, specific genotype many times and then, in turn, expose that replicated genotype to as diverse an array of environments as possible. However, the array of phenotypes that developed through this procedure would *at best* provide only an estimate of the norm of reaction for that one genotype, because in reality it would take an infinite number of genotype replicates exposed to an infinite number of sets of environmental conditions to determine the exact norm of reaction.

Accordingly, to suggest that groups differ in any necessary way because of a range of behaviors associated with their genotypes is to ignore the fact that the actual norm of reaction, for *any* living organism, cannot be known theoretically or in practicality—simply, by definition we cannot reach infinity. Moreover, any argument that there are "average expected environments" within which genotypes may be expressed, and that somehow the presence of such settings represents a necessary constraint on behavioral development, is also ill-formulated. Both the concept of norm of reaction and the data supporting the fusions illustrated in Figure 2 underscore the point that, as the context changes in new ways, we may expect that the outcomes of the integrated processes of human development will also change.

McGuire and Hirsch have drawn a compatible conclusion about genotype-environment fusions.

> There are undeniable genetic differences between individuals in anatomical, physiological, and behavioral traits. The genetic differences are important in determining how any individual develops in a given environment. Observation of development in one environment however, provides no basis for predicting how the same individual might have developed in a different environment. In fact, it does not give 100% predictability of how a replicate with an identical genotype might develop in the same environment.[36]

In sum, then, evidence of gene-environment integrations denies the usefulness of a genetic determinist view of behavior development, whether expressed in an extreme or a weakened form. The plasticity of behavioral development associated with these fusions exists, McGuire and Hirsch note, because:

> Any phenotype is the result of the interactions of the alleles at all loci, the interactions within and between loci, and the interaction of this genotype with environmental variables. A phenotype cannot legitimately be discussed independently of the environment in which it has developed. Similarly, environmental variables (teaching techniques) cannot legitimately be discussed independently of the genotypes they were tested on.[37]

Thus, McGuire and Hirsch note that plasticity exists because a genotype cannot be discussed appropriately independent of the context with which it is integrated *and* because the context cannot be understood adequately independent of the genetic characteristics of the organisms with which it fused. In other words, similar plasticity can be seen also when one turns the information displayed in Figure 2 around a bit and considers how a given environment may be dynamically integrated with organisms differing in their heredity (see Figure 3).

Two views of the relationship between nature and nurture levels are also illustrated in Figure 3. In part (a) of Figure 3, I depict the position cultural determinists take (discussed in Chapter 1). Here the contribution of environment (for instance, societal institutions, culture) to behavior is depicted as being "direct." Thus, a particular environmental event or set of events (E_1) is seen as leading directly to a particular behavioral outcome (B_1), and any particular biological features of the person are not considered seriously or systematically. The belief here is one of complete malleability, of absolute plasticity.

However, as with the former argument regarding nature contributions, this view is not tenable. Just as the influence of heredity on behavior is contingent on the characteristics of the environment with which it is fused, environmental contributions to behavior are contingent on characteristics of the biological levels of the organism. From this view, the same environmental event (for example, contraction of a disease or exposure to a particularly stimulating lecture) or group of events (for example, events associated with middle-class as opposed to upper-class membership) will lead to different behaviors depending on the integrations that occur with the characteristics of the biological levels of the person. This view is illustrated in Figure 3 (b), where the same environment (E_1) can be associated with an

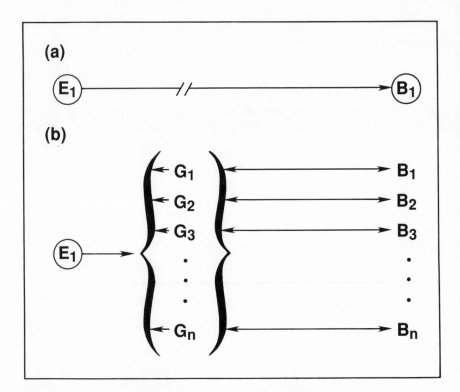

Figure 3.   (a) Environment (E) does not directly lead to behavior (B). (b) Rather, the influence of environmental levels will vary when integrated with (1) people having genetic-level (G) characteristics that are different and (2) the behavioral level.

array of behaviors (B_1 to B_n) as a consequence of dynamic integrations with people having different heredities (G_1 to G_n). Another basis of relative plasticity in development is thus promoted.

This plasticity may be illustrated in several ways. Consider a very general set of experiences: those associated with being a child of upper-middle-class parents. Imagine that such parents had two children who were dizygotic twins (fraternal twins). Such siblings are born of the same pregnancy but from two separate ova that are fertilized at the same time, so although born together, and unlike monozygotic twins, they have different genotypes. If one of these twins was born with Down's syndrome while the other was born with a normal complement of genes, a situation would result wherein children born of the same parents at the same time would potentially be exposed to the "same" environmental events. But, whatever the experiences encountered by the Down's syndrome twin, the effects of those experiences could not be expected to result in behaviors falling within a range exactly identical to that of the sibling. For instance, despite the advances in special education noted earlier, one still cannot expect a Down's syndrome child to attain the same vocation and live as long a life as the sibling born with the normal genotype. Thus, the hereditary nature of the organism imposes limits on the possible contributions of environment.

There are other illustrations of such dynamic integrations. For instance, if a mother contracted rubella during pregnancy, adverse physical and functional outcomes for the infant might follow. But, this same experience (contraction of rubella) may or may not lead to these outcomes, depending on when in the development of the person this disease is contracted. If the experience occurs during the embryological period, these negative effects are likely to occur; if it happens in the late fetal period, they are not likely to happen. Similarly, excessive maternal stress will or will not be more likely to lead to certain physical deformities (such as cleft palate) depending on the developmental time during which it occurs. Thus, here again, the biological features of the developing person moderate the influence of experience on his or her development.

It may be concluded, then, that even if one is talking about very narrow sorts of environmental experiences (such as encountering a specific person in a specific transitory situation—for example, a rude passenger on an elevator) or very broad types of experiences (such as experiences associated with membership in one culture versus another), the influences of these environments would not be the same if they were integrated within hereditarily (genotypically) different people. And even if it were possible to ensure that the different people had identical experiences, the influences would not be the same. As long as the nature (biology) of the person is different, the contributions of experience will vary.

Because of genotype uniqueness, all individuals will interact with their environments (whether the same or different) in unique, specific ways. So the environment always contributes to behavior, but the precise direction and outcome of this influence can be understood only in the context of the genetic individuality of the person. In turn, individual differences in genetic makeup do not in and of themselves directly shape behavior. Integration with an environment, itself having a host of distinctly individual features at each of several different levels of organization, must be taken into account.

However, if the trillions of possible distinct genotypes are difficult to envision, it is no less formidable to conceptualize the potential complexity and diversity of the context of human life—which in order to appreciate the full meaning of the developmental contextual perspective is a necessary task.

THE RELATIONSHIP BETWEEN BIOLOGY AND THE CONTEXTS OF HUMAN LIFE

The core idea in developmental contextualism is that biology and context cannot be separated, that both are fused across life. One way to begin to illustrate just what is involved in this relationship, even for one person, is to consider the diagram presented in Figure 4, which is drawn from a developmental view of a child and, most immediately, one of his or her parents. In the figure the circle on the center-left represents an individual child. Within the child are represented (as slices of a pie) several of the dimensions or levels of the child's individuality: his or her health, physical and biological status, personality and temperament, behaviors, and attitudes, values, and expectations. All these dimensions, as well as others (included as "etc."), are represented as existing in a manner contingent on the child's development level, which is also a pie slice within the child figure. In short, what is represented by this portion of Figure 4 is that a child is a complex and differentiated—multilevel—biological, psychological, and behavioral organism and that this differentiation is not static, but rather develops across time.

However, just as genes are embedded within a differentiated, multilevel environment, so too are children. Accordingly, I have used a circle on the center-right of Figure 4 to represent a parent of the child: the parent also is a complex and differentiated (multilevel) biological, psychological, and behavioral organism and is, as well, a person who is developing across his or her life.

The mutual influence between child and parent, their fusion with each other, is represented in the figure by the bidirectional arrows between them. It is important to point out that we may speak of dynamic interactions between parent and child which pertain to either *social* or *physical* (e.g., biological or physiological) level integrations. For example, in regard to

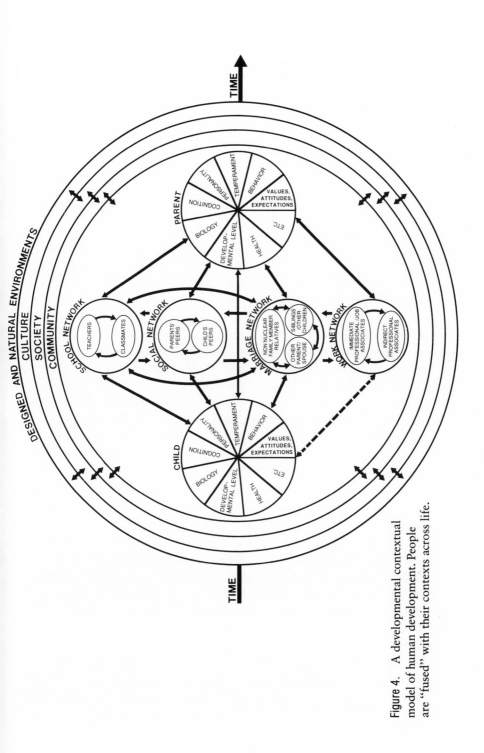

Figure 4. A developmental contextual model of human development. People are "fused" with their contexts across life.

social relationships, the parent's attitudes, values, and expectations might involve demands that the child pay attention, as when the parent directs the child to pick up toys, do homework, or go to bed. If the child does not pay attention, this lack of consistency with parental expectations might light the parent's "short fuse" for tolerance; the parent then scolds the child, who then cries, creating remorse in the parent and eliciting soothing behaviors; the child then becomes calm, snuggles up to the parent, and now both parties in the relationship show positive emotions and are happy.

The dynamic interactions also involve biological-level or physiological-level processes. For example, parental religious practices, rearing practices, or financial status may influence the child's diet and nutritional status, health, and medical care. An infectious disease contracted by either parent or child can also infect the other member of the relationship. And the health and physical status of the child influences the parent's own feelings of well-being and the parent's hopes and aspirations regarding the child.

Thus, the child's internal, physiological status and development are not disconnected from his or her outer, behavioral and social context (in this example, parental) functioning and development. The inner and outer worlds of the child are, then, fused and dynamically interactive. The same may be said of the parent and of the parent-child relationship. Each of these focuses—child, parent, or relationship—is part of a larger, enmeshed *system* of fused relationships among the multiple levels that compose the ecology of human life.[38]

For instance, illustrated in Figure 4 is the idea that both parent and child are embedded in a broader social network and that each person has reciprocal reactions with that network too. This set of relationships occurs because both the child and the parent are much more than just people playing only one role in life. Focusing on the sorts of networks seen in contemporary Western, industrialized societies, the figure illustrates that the child may also be a sibling, a peer, and a student; the parent may also be a spouse, a worker, a peer, and an adult child (of the child's grandparents; represented in the figure by the "nonnuclear family member" component of the marriage network). All these networks of relations are embedded also within a particular community, society, and culture. And finally, all these relations are continually changing across time, across history. Simply, for all portions of the system of integrated biological-environmental level relationships envisioned in developmental contextualism, whether they involve instances of Western, Eastern, or Third World individual and social context characteristics, change across time is an integral, indeed inescapable, feature.

Thus, Figure 4 is only one way to represent the integrated and multiple person-context relationships that characterize life across the span of human development. Given the complexity of Figure 4, and the fact that analogous figures representing people embedded in different cultural and historical contexts would be similarly complicated, a developmental contextual perspective may seem unnecessarily convoluted and complex. But just the opposite is the case.

The study of human development cannot rely on models that distort the character of human life through reliance on overly simplistic genetic and/or environmental models. The perspective and information presented in this chapter underscore the view that diversity of peoples and of contexts is the rule in human development. Global approaches that ignore or minimize those differences do not make for good science. In addition, they fail to provide an adequate basis for the design of policies or programs that will serve the specific needs of individual people and groups living in their particular families, communities, and societies. In short, if we are to do "good science" and provide "good service," our theoretical models must be able to engage the complexity and diversity of human life.

PLASTICITY AND THE POTENTIAL FOR CHANGE

The ubiquitous presence of change—what we have termed the potential for relative plasticity in human life—is the key difference between the doctrine of biological determinism and the developmental contextual perspective. It is a difference between pessimism and optimism. It is a difference between believing that we must accept and live with our biology, with the purportedly fixed and immutable limits it imposes, or acknowledging and cultivating (and celebrating) the potential for change and improvement that our biology offers.

Ultimately, then, the distinction between biological determinism and developmental contextualism is, respectively, whether society takes actions that attempt to make the best of the purportedly rigid hand dealt to us through our genes by evolution, or whether society pursues policies that attempt to capitalize on the potential for change inherent in each person and group and thereby enhance the human condition. Do we, as humane biological determinists, devise means to allow people to live as best they can within their biological limitations, which are imposed on them because they have genes that make them a Black or a woman, for instance? Or, as humane developmental contextualists, do we attempt to conceive of, and implement, means through which each person—no matter what his or her racial or sexual group may be—is given the resources to actualize his or her own potential for change?

As a developmental contextualist, my answer to the last question is yes. In making such a response, however, I am aware that I have imposed on myself the challenge to show *how* one can foster actualization of individual potential. How can individuals act as agents in promoting their own enhanced development? How can social policies be directed to promote such actions? Chapter 7 provides my answers to these questions.

7

DEVELOPMENTAL CONTEXTUALISM AND SOCIAL POLICY: THE ENHANCEMENT OF HUMAN LIFE

It is as heavy a responsibility to inform man about aggressive tendencies assumed to be present on an inborn basis as it is to inform him about "original sin.". . . A corollary risk is advising societies to base their programs of social training on attempts to inhibit hypothetical innate aggressions, instead of continuing positive measures for constructive behavior.

—T. C. Schneirla,
reviewing Lorenz's On Aggression

Philosophical ideas about the nature and nurture of human development influence much more than scientific theory and research. Almost every person has some conception, explicit or abstract, about what human life is and how it develops. Almost every person has some belief about where the essence of humanity derives from, be it through "things" placed into humans—by God, by evolution, or by genetic inheritance—or through the things into which humans are placed, such as the family, community, society, or culture.

The events and policies in Nazi Germany that applied a biological determinist view of human life demonstrate how beliefs about the nature and nurture of human development may influence government laws and social policies. I have also indicated how such beliefs may affect educational practices, paths of professional and vocational opportunity, general social intercourse, and individual psychological functioning and behavior. Beliefs about the nature and nurture of human development influence as well our values about and attitudes toward human diversity, and about the basis, meaning, and value of individual differences among people. Thus, our views about nature and nurture influence our opinions about why certain people may have problems and about what, if anything, can be done for remediation. Finally, our views about nature and nurture may influence how we behave toward those who differ from us. Do we avoid them or reject them?

Do we, de facto if not de jure, keep them apart from us? Or does our behavior welcome diversity and celebrate the richness of human behavior and potential that human differences teach us?

Answers to such questions have had and will continue to have an influence on the social policies promoted in society. Because of this continuing and potentially critical role, this book has focused on two distinct approaches to addressing the nature-nurture issue and thus to formulating policy answers to such questions: (1) the doctrine of biological, or genetic, determinism and (2) developmental contextualism.

Konrad Lorenz, both during and after the Nazi era, and contemporary sociobiologists, such as Melvin Konner, assume that human life is biologically constrained to manifest aggression; in addition, such scholars assume that xenophobia, militarism, ruthless selfishness, and other less-than-desirable characteristics are necessary and ubiquitous features of human life. Adherence to the doctrine of biological determinism leads them to contend that, however society may lament it, such characteristics are carried as blueprints in our genes and are the outcome of our evolutionary past.

For biological determinists the social policy implications of this situation are clear. Because society will always be faced with members engaging in militaristic aggression, in out-group prejudice, and in ruthless exploitation of one group by another, people must organize social institutions to make the best of the bad situation afforded them by their "human nature"—for instance, as Lorenz suggested, society should try to channel the innate (biologically fixed) militant enthusiasm of youth into constructive endeavors.[1] Biological determinists believe that we must recognize the savage nature evolution has given us, accept the immutable character of our genes, and try as best we can to counteract our innate evil—the "original sin" of which T. C. Schneirla spoke.

My presentation of the developmental contextual perspective (in Chapter 6) indicates that the pessimistic approach to social policy found in biological determinism is not at all warranted. In fact, the evidence for fusions among biological (e.g., genetic) and environmental levels and for the relative plasticity in human life which derives from the dynamic integrations among levels suggests that a considerably more optimistic approach to policy is appropriate. Policies should be aimed at enabling individuals to realize their potential for positive growth—for constructive change, as Schneirla puts it—in full recognition that such growth is possible. From a developmental contextual perspective, then, policies should be formulated to allow individuals to serve as agents in fostering their own constructive change.

Can this be done? Figure 4 illustrates how individuals are fused with their settings across the development of their lives and are integrated with a complex multilevel and changing context. It is through this fusion that people—as dynamically interactive *with* biology and context—can act to advance their own development, rather than being constrained by their biology (or context).

HOW INDIVIDUALS CAN AFFECT
THEIR OWN DEVELOPMENT

Relationships between any given individual and other persons in his or her social world differ from person to person, because each of us is distinct. We all have characteristics of biological, psychological, and behavioral uniqueness which are associated with our own particular history of biology-context fusion.[2] Each of us has our own particular body build, facial characteristics, personality, temperament, and beliefs, values, and opinions.

Because of your individual characteristics, you will elicit somewhat different behaviors in the people with whom you interact. The different reactions of people act as feedback for you, and that feedback is distinct because it depends on how your particular physical or behavioral characteristics affected other people. As a result, the feedback becomes part of your ongoing history of unique person-context fusions. The feedback you receive will promote your further development as a singular individual. Because this distinct feedback arises at least in part because of your already established individuality, you in this manner act as a producer of your own continued development. This ongoing process of person-context feedback provides the basis for a *circular function* between you and your environment.[3] Such circular functions allow you to be an active agent in your own development, and this circular process allows scientists to characterize the processes involved in your development as relatively plastic in character.

The reciprocal person-context relations, which across our individual life-spans allow us to be active agents in our own development, appear to have been involved also in human evolution. In other words, the circular (person-context) functions in which all human beings engage may have acted, across history, to be a source of our own evolutionary development. Lewontin and Levins state: "The activity of the organism sets the stage for its own evolution. . . . The labor process by which the human ancestors modified natural objects to make them suitable for human use was itself the unique feature of the way of life that directed selection on the hand, larynx, and brain in a positive feedback that transformed the species, its environment, and its mode of interaction with nature."[4]

Similarly, C. O. Lovejoy argues that the social relationships that led to brain evolution were themselves altered when larger-brained, more plastic organisms were involved in them; in turn, new social patterns may have extended humans' adaptational pressures and opportunities into other arenas, fostering further changes in brain, in social embeddedness, and so forth.[5] Donald Johanson and Maitland Edey describe Lovejoy's position as one that requires examination of

> the mechanism of a complex feedback loop—in which several ele-
> ments interact for mutual reinforcement. . . . If parental care is a
> good thing, it will be selected for by the likelihood that the better
> mothers will be more apt to bring up children, and thus intensify any
> genetic tendency that exists in the population toward being better
> mothers. But increased parental care requires other things along with
> it. It requires a greater IQ on the part of the mother; she cannot
> increase parental care if she is not intellectually up to it. That means
> brain development—not only for the mother, but for the infant
> daughter too, for someday she will become a mother.[6]

Johanson and Edey explain that the feedback loop involved in the evolution of primates was not merely a bidirectional one, involving the influence of A on B and, in turn, of B on A. To the contrary, multiple sources of influence were involved, each integrated with all the others like the dynamic relationships in the multiple levels of organization fused in gene-environment relations, described in Chapter 6. Therefore, I would expand the views of Johanson and Edey about the brain development of mothers and infant daughters to include both parents and children of both sexes and to include morphological characteristics other than the brain (for example, changes in the hand and in the larynx, stressed by Lewontin and Levins[7]), and I would broaden the context fused with these changes to include the entire family, the larger social group or system, and activities of people beyond those involved with child care—for instance, work. Johanson and Edey's views appear to be quite consistent with this expansion. They note:

If an infant is to have a large brain, it must be given time to learn to use that brain before it has to face the world on its own. That means a long childhood. The best way to learn during childhood is to play. That means playmates, which, in turn, means a group social system that provides them. But if one is to function in such a group, one must learn acceptable social behavior. One can learn that properly only if one is intelligent. Therefore social behavior ends up being linked with IQ (a loop back), with extended childhood (another loop), and finally with the energy investment and the parental care system which provide a brain capable of that IQ, and the entire feedback loop is complete.

All parts of the feedback system are cross-connected. For example: If one is living in a group, the time spent finding food, being aware of predators and finding a mate can all be reduced by the very fact that one is in a group. As a consequence, more time can be spent on parental care (one loop), on play (another) and on social activity (another), all of which enhance intelligence (another) and result ultimately in fewer offspring (still another). The complete loop shows all poles connected to all others.[8]

This system of reciprocal integrations across levels of organization, this "complete loop," is shown in Figure 5, which illustrates that the bases of the plasticity of human beings evolved in a complex system of fused, or dynamic interactional, relationships among characteristics on the social, individual, and biological (e.g., morphological) levels. Accordingly, there is substantial reason to believe that the ways in which we as individuals act as agents in our own development were central as well in the history of our evolutionary change. Thus, rather than passively and mechanistically acting according to the social roles specified by our genes—functions envisioned in the Social Darwinism of biological determinism—we as individuals actually shape our lives and the social institutions within which we live. Our discussion of such processes can focus on three distinct though interrelated ways in which individuals not only shape the feedback they receive but also, in so doing, contribute to their own context and their own ensuing development.

PEOPLE INFLUENCE PEOPLE WHO INFLUENCE THEM

For at least two decades some developmental psychologists and psychiatrists have been stressing that people can influence those who influence them[9]—

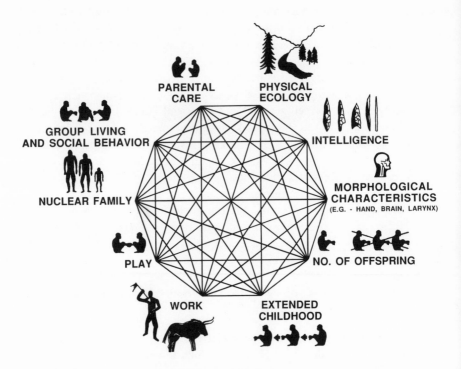

Figure 5.   Components of the system of individual-level and contextual-level integrations believed to be involved in the evolution of human functioning. Adapted from D. C. Johanson and M. A. Edey, *Lucy: The Beginnings of Humankind* (New York: Simon & Schuster, 1981).

that is, that development involves reciprocal relations rather than only unidirectional (for example, from parent to child) relations.[10] Furthermore, as explained above in regard to the concept of "circular functions," people get feedback as a consequence of their influences on others.

Such circular functions arise in relation to characteristics of physical and behavioral individuality. Children who differ in physical characteristics relating to their body build or physical attractiveness, or to whether they were born prematurely, receive different reactions from parents, teachers, or friends—for example, physically attractive and/or average-build children are judged by adults and other children in terms that are much more positive than is the case with unattractive and/or chubby children.[11] Social feedback is consistent with such reactions. Other children want to come closer to (for example, when playing with) average-build children than to chubby children, for instance.[12] In turn, premature infants elicit less parental involvement during the newborn period than their full-term counterparts (perhaps because of the fragility of premature infants), but premature infants receive more attention across their first eight to nine months, probably because of the infants' inability to initiate or maintain behavioral interactions with their parents, and the parents' desire to have such interactions.[13]

The subsequent development of such children appears to be consistent with the different feedback they receive. For example, physically unattractive children have more behavioral and adjustment problems than do physically attractive children,[14] and premature infants, probably because of the extra stress they may generate for already highly stressed parents, appear to run an increased risk of being mistreated.[15]

Similar circular functions in regard to characteristics of behavioral individuality are also apparent. Children who differ in temperament put differing caregiving demands on their parents. For instance, children who are biologically arrhythmic (e.g., who sleep and wake or get hungry at unpredictable times), who are intense and negative, who tend to withdraw from new foods or situations, or who are slow to adapt to any new features of their context (e.g., a change in schedule or routine) are more difficult to deal with than children who are rhythmic, who tend to approach and adapt easily to new situations and who react with moderate intensity and positive moods.[16] Children with temperamental differences also behave differently with their parents and other adults,[17] are regarded in different ways by their caregivers (e.g., teachers),[18] and have a different probability of developing a problem requiring clinical (e.g., psychiatric) intervention.[19] In sum, temperamentally difficult children are more likely than temperamentally easy children to have personal and social problems with family, friends, and other people in their social worlds.

PEOPLE PROCESS THEIR WORLD IN UNIQUE WAYS

One assumption on which the field of human development is based is that people do not remain the same over the course of infancy, childhood, adolescence, and so on. In the typical case, physical, cognitive, social, and emotional characteristics undergo systematic changes. The history of the field is witness to a vast array of scientific information that supports this assumption.

Therefore, the fact that you, as an individual, undergo developmental changes means that you are, in effect, a somewhat different person at various points in your life span. More interesting, such development means that the same experience—for instance, the same parental childrearing strategies (e.g., being "grounded" for misbehavior) occurring or implemented at different points in your life (late childhood versus mid-adolescence)—may be treated (processed) by you differently, and because you processed it differently the same experience could have had differing effects on your behavior.

To illustrate, children in a period of rapid growth may respond to a physical-education program differently from children in a period of slow or little growth. Younger children will be less able to use abstract rules for self-control than cognitively more advanced children.[20] Similarly, children who are in the process of establishing secure social relationships with their parents (e.g., children during their first year of life) are likely to have sets of social skills that are different from those of children who have already established such attachments.

PEOPLE SELECT AND SHAPE THEIR CONTEXTS

A competent person is one who is capable of showing appropriate behavior and thought. To behave appropriately in a changing and/or complex world, one cannot always do the same thing. People need to be flexible about how they behave and how they think about the issues and problems encountered in life.[21] Flexibility, made possible by the relative plasticity of the developing person, allows a person to change to meet new problems or demands *or* to change the context in order to better fit personal objectives.[22] A competent person can effectively modify his or her own behavior *and/or* the features of the social situation in which he or she is engaged. We can, for instance, change the topic of conversation if we find we are boring or upsetting others; or if we are bored or upset by what is being said, we can turn the conversation around to more pleasant topics.

To indicate how such efficient self-regulation is a way you can act as a producer of your own development, it is useful to look at what has been termed the "goodness of fit" model of relations between people and their contexts.[23] Just as you bring your particular individual characteristics to a given social setting, the social and physical components of the setting place certain demands on you. These demands may take the form of (1) attitudes, values, or stereotyped beliefs held by others in the context regarding your particular physical or behavioral characteristics; (2) the attributes (usually behavioral) of others in the context, with whom you must coordinate, or fit, your characteristics in order for good interactions to exist; or (3) the physical characteristics of a setting (for example, the presence or absence of access ramps for the handicapped) that require you to possess certain characteristics for the most efficient interaction within the setting to occur. Your individuality in differently meeting these demands provides a basis for the feedback you get from the people in your social world. Psychiatrists Alexander Thomas and Stella Chess and psychologists Jacqueline Lerner and Richard Lerner indicate that effective psychological and social functioning is most likely to occur when your characteristics of individuality match (or "fit") the demands of a particular setting.[24]

But what are the precise competencies required to attain a good fit within and across time? You must be able to evaluate appropriately (1) the demands of a particular context (e.g., you must ask "What is required or expected of me?"); (2) your psychological and behavioral characteristics (e.g., you must ask "Can I do—physically or behaviorally—what is expected of me?"); and (3) the degree of match that exists between the first two questions—that is, the "fit" between your answers to the first two questions. In addition, other cognitive and behavioral skills are necessary. You must be able to select and gain access to contexts with which there is a high probability of match and to avoid contexts where poor fit is likely. For instance, you may want to avoid playing tackle football or rugby if you are five feet tall, weigh ninety-seven pounds, and have frail bones. You might seek out settings (e.g., small, informal parties) wherein your particular skills (e.g., conversations about old movies, world events, etc.) would be welcome. In addition, in contexts that are not usually selected—for example, your family of origin or your assigned work role—you must either have the knowledge and skills necessary to change yourself to fit the demands of the setting or you must alter the context to fit your individual attributes better.[25] To use the example of upsetting or pleasant conversations, you must decide whether good fit is more likely if you change what you contribute to the discussion or if you try to change the topic of conversation. Moreover, in most contexts there will be multiple types of demands, and not all of them will have identical requirements. You must therefore be able to detect and evaluate such complexity and determine which demand to meet when not all can be met.

As you develop competency in self-regulation, you will become a more active selector and shaper of the contexts within which you develop[26] and increasingly be able to affect others (your friends, parents, spouse, co-workers) at least as much as they influence you. Through such influences you will develop your capacity to be an active contributor to your own development.

Individuals can foster their own development in many ways. Rather than being a product of their biology, a fixed and immutable outcome of prescriptions in their genes, people are an active, contributory part of the system of fused integrations among biological and contextual levels which promote their particular path of behavioral development. Counter to the view of sociobiologists that the ultimate rationale for our existence is to reproduce our genes and that we are nothing more than reactive, indeed robotic,[27] machines controlled by such biological directives, the developmental contextual perspective suggests that all features of our biology are enmeshed in a system of dynamic interactions (integrations) with the multilevel context of life. In fact, the attainment of reproductive maturity in women is a powerful example of just how fused with the context one key feature of human reproduction is. It is useful to consider here how this key feature of individual biology is interwoven thoroughly with the social context.

FEMALE REPRODUCTIVE MATURITY: AN EXAMPLE OF CONTEXT AFFECTING BIOLOGY

A developmental contextual view of the role of biology in human development differs in several ways from a biological determinist conception of this role. First and foremost, genes do not represent fixed and immutable blueprints for behavior and social standing. Genes are plastic entities that are fused with a complex context. Given this fusion, there is not a one-to-one relationship between genotype and behavior. In dynamic integration with the specific context within which it is embedded, any genotype can be associated with a broad range of distinct behaviors, often labeled the "norm of reaction" (see Chapter 6).[28]

Second, because the plasticity of genes and of the individual and social processes in which they are involved arises from the fusion of organisms and contexts, people themselves contribute actively to their own plasticity, to their own development. The dynamic integrations between people and their contexts are ubiquitous parts of the human life span. Moreover, the role of the person in shaping his or her own context appears to have been a key component of human evolution.

Therefore, third, a human being's position in life, his or her place in the social hierarchy, is not a predetermined, genetically forced outcome of natural selection. Quite to the contrary, people's roles in their world must be understood in connection with the historical and social conditions of the world within which they transact.[29] Simply, "in a world in which such complex developmental interactions are always occurring, history becomes of paramount importance."[30]

To illustrate, even a physiological event tied as closely to reproduction as attaining reproductive maturity is related to historical changes. For instance, the age at which women reach menarche has declined across history in different countries. Among European samples of youth, there was a decrease of about four months per decade from about 1840 to about 1950, and the rate seems to have slowed but not stopped.[31] Within samples in the United States, however, the decline in age of menarche seems to have stopped about 1940; since that time, 12.5 years has been the expected value for menarche among White middle-class Americans.[32] The most dramatic secular trend was evidenced in Japan. From the immediate post–World War II years until about 1975, there was a decline in the average age of menarche of eleven months a decade.[33]

Such changes in timing of reproductive maturity, termed a secular trend, are but one instance of significant physical changes occurring in males and females over even relatively recent history. Since 1900, for instance, children of preschool age have been taller on an average of 1.0 centimeter and heavier on an average of 0.5 kilogram per decade.[34] In addition, during the growth spurt that occurs as part of the pubertal changes of adolescence, adolescents have shown increasingly greater increases in height (2.5 centimeters) and weight (2.5 kilograms) across this recent historical period.[35]

These physical and physiological changes have occurred too rapidly to be due to genetic changes across time. Instead, current evidence points to the role of historical changes in nutritional regimens and resources and in health care practices as the basis for these changes.[36] Thus, the presence of such historical changes in attributes as central to the argument of sociobiologists as are reproductive and physical maturity ones, and the apparent bases of such historical effects, make claims that evolutionarily based, genetic differences between races are the key source of differences in sexual maturity rates implausible.

IMPLICATIONS FOR SOCIAL POLICY

How important for the role of science in social policy is the fact that characteristics of an individual's reproductive functioning and growth are dynamically fused with the social and historical context? What are the implications of the evidence that people are active agents of their own relatively plastic development?

One immediate implication is clear: any policies based on the speculations of J. Philippe Rushton about evolutionarily based and genetically fixed differences in the reproductive strategies of Blacks, Whites, and Asians would be scientifically ill-advised. Indeed, on the basis of the evidence we have reviewed, Rushton's assertions that the earlier sexual maturation rate of Blacks is a product of their more r-like reproduction strategy[37] appears to be egregiously flawed. Moreover, the fact that both socioeconomic and cultural differences, especially those tied to certain nutritional and health variables, are known to moderate pubertal maturation differences among current racial groups adds yet another point in criticism of Rushton's argument. For instance, while the median age of menarche for Cuban Blacks or Whites is 12.4 years, the lowest age listed in a survey of ethnic differences in growth and development, the median age of menarche for the Black Bundi tribes of New Guinea is 18.8 years, the highest found in the same survey.[38] Thus, groups of Blacks are at both the highest *and* the lowest portions of this distribution.

Reproductive maturity is therefore not just an attribute of human functioning driven by evolutionarily shaped genes. Instead, it and other physical variables—and personality, cognitive, and social behavior variables as well[39]—are moderated by sociocultural factors changing across history. Given, then, the mutuality of plasticity versus constraints inherent in developmental contextualism, such historical (temporal) embeddedness means that:

1. While people are striving to change and/or limit their contexts, their contexts are simultaneously acting to change and/or limit them.

2. As people and/or their contexts change (as they will across developmental and historical time, respectively), the outcomes of people-context transactions can change as well. New behaviors and/or different social roles may therefore emerge quite readily.

So we cannot tell what history will bring. Changes can be for the worse, but they can be for the better as well. We need not be pessimistic, therefore, that changing the social order—introducing humanitarian policies, for example—will either destroy "the Best" of society or be a fruitless, self-defeating waste of time.

Such pessimism, however, is precisely what characterizes the claims of Nazi ideologues, sociobiologists, and other biological determinists. For instance, E. O. Wilson says: "Even in the most free and most egalitarian of future societies . . . men are likely to continue to play a disproportionate role in political life, business and science."[40] And similarly, psychologist Arthur Jensen has said that differences in social status, while perhaps not fair in the philosophical sense, are "nature's way"—the outcomes of natural selection processes. He contends: "We have to face it, the assortment of persons into occupational roles is not 'fair' in any absolute sense. The best we can hope for is that true merit, given equality of opportunity, [will] act as a basis for the natural assorting power."[41]

Indeed, from this view we dare not interfere with the natural selection process lest we create even *more* social differences between people. Exemplifying this Social Darwinist "warning" are the views of psychologist Richard Herrnstein, who argued:

> The privileged classes of the past were probably not much superior biologically to the downtrodden, which is why revolution had a fair chance of success. By removing artificial barriers between classes, society has encouraged the creation of biological barriers. When people can take their natural level in society the upper classes will, by definition, have greater capacity than the lower.[42]

To Herrnstein, then, our present society is totally "just and fair," and consequently the poor, weak, homeless, and unemployed are in their proper, biologically established place. This is so, in Herrnstein's view, because no "artificial" (societally constructed) barriers have blocked people from falling into their biologically proper role. There have, again in his view, been no racial, ethnic, or gender prejudices; there have been no discriminatory laws or social practices; there has been only unimpeded social opportunity.

Indeed, those adopting a position like Herrnstein's might argue that because there are now no artificial barriers the only laws and/or social programs that should be enacted are those that facilitate freedom of opportunity.[43] Ensuring that all people have equal opportunity in society will facilitate their entering into the positions and statuses for which they are biologically predisposed.

However, those taking a position like Herrnstein's might argue also that such laws should not go so far as to guarantee that some opportunities must be reserved for people of a certain group. For instance, such laws should not say that because a given group (e.g., women) have not had a fair opportunity to enter a given profession (e.g., medicine), one should correct this past "imbalance" by specifying a percentage (e.g., a quota of, say, 20 percent) of places that should be reserved for this group. Such a quota system would only distort the "natural assorting power" Jensen spoke of[44] and put people into roles they were not biologically fit to play. Indeed, to people taking a position like Herrnstein's, the very history of individuals not engaging in such roles is proof of their lack of biologically based ability, not of past discrimination.[45] The argument of Lenz exemplifies this reasoning. Concerning women's intellectual and creative achievements, Lenz argued that "although female students may do well on university exams, their development is inevitably arrested shortly thereafter"[46] and therefore " 'great women' endowed with 'greatness' in the sense of outstanding creative faculty are practically unknown."[47]

It seems, then, that in continuity with the social policy prescriptions made by biological determinists of the Nazi-era, contemporary biological, or genetic, determinists have advanced markedly parallel policy recommendations. Moreover, as I noted above, such appeals are continuing and are among the ideas vying for the attention and support of the makers of public policy today. For instance, and redolent of Rushton's 1989 belief that his theories of the biological basis of race differences represent a "truer picture of human nature,"[48] we may recall Kachigan's 1990 appeal for the institution of a sociobiologically informed "natural biological order among humans" (see Chapter 4).[49] Indeed, Kachigan's call is presented in an urgent tone because of his belief, recalling the fears expressed by Ploetz, Schallmayer, Binding, and Hoche,[50] "that when the least desirable parents bear more children than the most desirable parents, the average of the population actually *deteriorates* from one generation to the next."[51]

Kachigan's concerns may appear to be an implicit call for institution of a sociobiologically based eugenics policy to "save the race." In some quarters, such ideas have even moved beyond mere implication and become explicit recommendations, even through the time of this writing, of those persuaded by genetic determinist ideas. Indeed, these appeals have been a "tradition" continuing for decades. For example, psychologist Arthur Jensen wrote in 1969:

> Is there a danger that current welfare policies, unaided by eugenic foresight, could lead to the genetic enslavement of a substantial segment of our population? The possible consequences of our failure seriously to study these questions may be viewed by future generations as our society's greatest injustice to Negro Americans.[52]

This concern found a parallel at the time in a position paper written by geneticists James Crow, James Neel, and Curt Stern for the National Academy of Sciences. Focusing on the topic of intelligence, they contended:

> A selection program to increase human intelligence (or whatever is measured by various kinds of "intelligence" tests) would almost certainly be successful in some measure. The same is probably true for other behavioral traits. The *rate* of increase would be somewhat unpredictable, but there is little doubt that there would be progress.[53]

In a related paper, Crow went on to write:

> I believe that we already know enough to predict that a selection program to increase IQ or g would work. There would be an increase, but the amount would be uncertain, because of uncertainty both in the true value of h^2 and in the assumption underlying its use as a predictor.[54]

Calling for a selection program to advance eugenics principles was in the 1960s an appeal for inappropriate policy based on poor science. It is no less so today. Yet Crow expressed a corresponding belief in the appropriateness of eugenics in his 1988 paper "Eugenics: Must It Be a Dirty Word?" and in 1989 the renowned psychologist and statistician Lloyd Humphreys advocated strongly that the United States government establish a national eugenics policy.[55]

In sum, then, the attempt to specify the social-policy implications of biological determinist ideas and to call for implementation of such policies is not just an issue of the past. It continues to be of concern in contemporary society. Clearly, however, these policy implications of past and present biological, or genetic, determinist, Social Darwinist thinking are not compatible with those implications that may be drawn from developmental contextualism. Developmental contextualism sees biology, in dynamic interaction with the context, as the liberator of human potential, not as an irrevocable constraint. Because of the relative plasticity of human development, we simply do not know what will result from altering person-context dynamic interactions. As such, it is my view that an experimental orientation to social policy is needed. Society must commit itself to testing and evaluating the range of possible outcomes that may arise when people are allowed to develop within altered social arenas and institutional settings. Such a commitment may not be a radical recommendation, since it is possible to construe the history of social policies enacted by society as (perhaps unacknowledged) "tests" of human plasticity.

That is, from the developmental contextual perspective I have been discussing we may view existing social policies, institutions, and arrangements as one set of "planned attempts" (interventions) to influence human life and development. Most, if not all, of these existing features of society have an impact on human life. For example, I and my colleagues Richard Birkel and Michael Smyer have suggested:

> Key elements of a national economic policy, for instance, the availability of financial credit, current interest rates, and the securability of desirable retirement packages, have important effects on the life-course decisions and plans made by large numbers of individuals. Deciding whether and when to go to college, to marry, to have children, to buy a home, to change jobs, and to retire are all influenced by economic policies consciously designed by lawmakers with intercohort and developmental impacts in mind. Thus, the developmental implications of a wide variety of policy interventions become particularly salient when viewed from the . . . perspective we espouse.[56]

Moreover, the history of alterations in enacted social policies over generations "—variation associated with cultural changes, with historical accidents, or resulting from conscious policy manipulation—[may be] viewed as [an important source] of information about the developmental impacts of interventions. The behavioral changes which may be linked to these changes provide instances of plasticity associated with particular person-context relations."[57]

Thus, it is possible to interpret the history of social-policy enactments in society as at least a partial test of how planned interventions may actualize human plasticity. But what we also need is a *commitment to evaluate* policy enactments as "experimental interventions" into human development and *theoretical guidance* for such a "social policy as experimental intervention" undertaking. Changes should not be instituted *only* because they represent variation from what has been done in the past. Instead, policymakers and social interventionists must commit themselves to offering a conceptualization of why particular person-context relationships should be altered, at specific levels of analysis, over a particular time span.

Only theories that integrate several levels of analysis across time, and that involve both the particular target of intervention *and* other people and social settings within which this target is enmeshed, will suffice. This more complex, multilevel theoretical model is necessary because, as illustrated in Figure 4, a key implication of developmental contextualism is that effects of interventions are not limited to a single target of change or to a single time within the life span when one may be acting. It is therefore useful to discuss some of the features of a developmental contextually guided approach to policy and intervention.

A DEVELOPMENTAL CONTEXTUAL APPROACH TO SOCIAL POLICY AND EXPERIMENTAL INTERVENTION

The fusion of people and their contexts, and the ubiquitous nature of change, provide the basis for believing in the relative plasticity of human development. In addition, fusion and change provide the basis for the approach to social policy and to intervention found in a developmental contextual perspective. There are several aspects of such an approach. A key idea that derives from this perspective is that one may attempt to alter human development by directing quite different interventions at any one of several different levels within the complex system of person-context fusions (as illustrated in Figure 4).

For instance, in regard to formulating policies associated with interventions to optimize intellectual functioning in the later years of life, the array of options available to promote intellectual activity and prevent cognitive decline among retirees may include counseling people about nutritional requirements; helping retired couples to develop joint interests; educating children and grandchildren about how they might contribute to a rich cognitive experience for their elders; helping communities develop meaningful volunteer experiences for older adults; expanding programs for hostels for the elderly; promoting positive social attitudes about employing retired people; encouraging corporations to focus on older adults in the research and development about, and marketing of, new technologies; and lobbying to reduce the tax rates for retired persons who decide to work.[58] From a perspective involving the sorts of dynamic, or fused, interactions emphasized in developmental contextualism, each of these interventions may help achieve the desired goal. Of course, a concerted and integrated effort involving multiple levels of organization may provide the most efficacious intervention.

Another implication of the concept of fusion is that, because individuals adjust to their specific social contexts, a fit, match, or congruence may actually exist in what may appear, to an outsider to the ecological setting, to be a disadvantageous condition.[59] When the perspective of the person one is trying to help is integrated into the array of information used by interventionists, a richer and possibly even more accurate understanding of the individual and the setting may be obtained. As a result, interventionists must be attuned to the logic and intelligence that exist in naturally occurring social situations. Therefore, interventionists should not proceed "as if" they alone "know best." They must be cautious, lest they alter people's lives in unintended ways. In short, interventionists should be conscious of the diversity of healthful and desirable developmental pathways that may exist and should promote such pluralism.[60]

The numerous potential routes to a successful intervention provide social planners with great flexibility. This flexibility is enhanced when one realizes that policies and interventions may affect not only those who are alive now but also future generations. Each of us is born at a particular moment of history. We share with other people born at that time a common set of experiences related to the particular social, political, and economic events we encounter. For instance, the 1920 "birth group" or "birth cohort" experienced the Great Depression in their childhood; members of the post–World War II "baby boom" experienced the Vietnam War during adolescence and young adulthood; and members of the 1980 birth cohort experienced a complete revision of the political "map" of Eastern Europe during their childhood years.

If a group experiences a set of events particular to their own time, they may develop attitudes or beliefs specific to that experience. Thus, it is possible to try to influence an entire cohort of people through envisioning policies and interventions that will introduce such variation into their lives. To illustrate:

> Some current intervention efforts in American society focus on teaching future generations to "just say no" to drugs, to resist and register complaints about uninvited sexual advances, and to become computer literate. Such interventions are designed to bring the hard-won experience of one generation to another, and to instill in future generations a set of attributes deemed desirable, if not essential.[61]

Another example is the possibility of intervention with future parents to prevent undesired outcomes in yet-to-be-conceived offspring—for instance, changing the type of birth control precautions sexually active young adolescents use in order to prevent the conception and birth of a child who will be at risk for several health problems because the mother is a young adolescent. In some cases, it may be more effective to direct interventions at generational targets other than those possessing the "problem" we want to correct. For instance, rather than trying to alter the behavior of an Alzheimer's disease patient it may be more efficient and successful to attempt to alter an adult offspring's skills in aiding the parent and/or to improve the son's or daughter's social support system. Indeed, because of the broad historical approach to intervention I am recommending, it may be useful in the case of Alzheimer's disease to try to alter social policies in the direction of enabling adult offspring to obtain the education and financial resources needed to provide this skilled care for their aged parents.

The example of attempting to change adult children in order to help them care for their aged parents better raises an important point that is of intergenerational, historical influence. Given the fusion of any policy enactment or intervention in a complex world composed of members of several birth cohorts, we may expect that a policy aimed at one generation may influence others as well. For instance, although one may institute an early education program (such as Head Start) in order to enhance the academic functioning of children later in their lives, such efforts might alter also parents' perceptions about their children's intellectual abilities, and indeed the entire family's commitment to education as a means to increase people's life chances.

In fact, these possibilities are confirmed by the results of a follow-up study of several samples of children, from low-income families, who had been enrolled in infant and preschool early childhood education programs.[62] At the time of the follow-up, when the children were between nine and nineteen years of age, they were (1) more likely to meet their schools' basic requirements, less likely to be assigned to special education classes, and less likely to be retained in a grade than were children in comparison (control) groups, and (2) more likely than children in the control groups to give achievement-related reasons for being proud of themselves. In addition, children in these programs did better than control-group children on standardized mental ability tests, and children who participated in the program had better nutritional practices and generally better health. Furthermore, and indicative of the intergenerational influence of this intervention, program participation altered children's familial context. Participation in the program influenced mothers' attitudes concerning school performance and also enhanced the mothers' vocational aspirations for their children.

THE SYNTHESES OF DEVELOPMENTAL CONTEXTUALISM

Developmental contextualism promotes policies and interventions that link the multiple, fused, historically changing levels involved in person-context relations across life. The developmental contextual perspective indicates that policies and interventions must be synthetic, not reductionistic. Only synthetic approaches can give policymakers or interventionists some idea about the broad effects their efforts may have—that is, the potential effects in contexts and among generations other than those targeted in a given intervention effort.

Of course, one cannot predict all outcomes. Changes in the world are probabilistic, and nonnormative events (for example, natural disasters, illnesses, accidents, social conflicts, and political upheavals) do occur.[63] So human life, fused with such occurrences, becomes probabilistic as well. Consequently, developmental contextualists planning interventions or policies must estimate the range of changes that may take place as a result of their efforts. They must recognize that the probabilistic character of human development and of the world with which humans are fused means that there may be several quite different results of a given intervention or social innovation.

Some social innovations may be detrimental; others may be quite positive. Most will probably have some negative and some positive features, depending on who one asks and how one evaluates a new program. Therefore, a cost-benefit analysis of new programs must be part of any policy or intervention. This cost-benefit assessment must involve also multiple contexts and generations. Given the complex, changing fusions between people and their contexts, it is necessary to assess programs with information from across these levels. And because the effects of programs in one context or among one generation may or may not be compatible with the effects at other levels, evaluations too must be integrated across levels.

For example, a psychological intervention aimed at changing a spouse's level of assertiveness might improve the self-esteem of the target person, but at the family level greater conflict may occur because the target spouse now introduces more variation into discussions of family decisions than before. In other words, a valued or desired change at one level of analysis may involve undesired changes at other levels. Accordingly, in planning intervention efforts, all levels of analysis that are theoretically reasonable to consider should be evaluated in regard to possible gains and risks.

In addition, because people's lives are embedded in history, such analyses of costs and benefits must have a long-term component.[64] We may not reap the fruits of new social programs until long after the program is introduced or until multiple generations of people have experienced the program. Moreover, equality of opportunity should be assured, and our social policies should also not hesitate to foster new numbers of people entering a social niche not previously occupied by others of their group. The exposure such social policies engender may result in a plasticity that can have unknown benefits for the group and for society.

Social policy is part of the sociocultural level of analysis that influences person-context relations (Figure 4). The political system within a society exists at a corresponding level of analysis, and of course the political system has an impact on the creation and/or deployment of social programs and policies. The relationship of programs and policies to politics introduces an additional source of plasticity and of constraints in development. For instance, Gilbert Steiner documents the inability of the federal administration to implement a coherent agenda of family-oriented policies due to the political pluralism of constituents.[65] In a pluralistic democracy, where value systems among subgroups are not always congruent, attempts to intervene

in human development through policy initiation may be thwarted by over-
arching political considerations.[66] And the political meaning of an intervention may vary across historical time:

> Compare the act of teaching basic reading and writing skills to young
> black children in America during the 1960s with the same activity in
> pre–Civil War America. It becomes apparent that efforts to enhance
> individual and group development may be readily viewed as civic
> responsibility, or as political insurrection. Indeed, the relative devel-
> opment of different groups in society is one of the most important
> and controversial issues of our age, and interventionists must not be
> naive about what is at stake.[67]

Thus, the "policy–program–politics" interrelationship is simply one more
source of variation in evaluating the risk-benefit ratio surrounding a policy
or intervention. The presence of this interrelationship underscores the idea
that developmental contextualists cannot separate themselves from the
people and from the social (including political) realities of the world they
study. If they attempt such separation, then (to paraphrase Eldridge Cleaver)
they will be not part of the process but part of the problem.

Interventionists and the formulators of social policy must integrate into
their work the perspectives of the people they are trying to help, must
function collaboratively with them. The perspectives of those people, and
the way they interpret their lives and settings, enhance the knowledge base
for programs and policies, and their validity. Developmental contextualism
involves integrations among multiple levels of organization. So that the best
possible programs can be designed, delivered, and evaluated, there must be
a bidirectional partnership between the individuals and groups at whom an
intervention is aimed, and professional interventionists and policymakers.
Collaboration between these two sets of partners may make it possible to
avoid replacing a dehumanizing biological determinism with an elitist and
equally oppressive environmentalist one.

From a developmental contextual perspective, the proper role of social
institutions and the social policies that shape them is to work with individuals to forge programs that allow them to maximize their plasticity—to
"test the limits" of their ability[68] and as much as possible to be active agents
in their own development.

A catch phrase in America in the 1960s was "Power to the people." From
a developmental contextual perspective, such power—to engage our social

world, to influence it, and to shape it in directions of benefit to us and to the others with whom we share it—does not have to be given to the people by some superordinate controllers of society. It resides naturally in us.

This is the heritage our evolution has given us, the meaning of our biology and our behavior. All of us, no matter what our race, religion, gender, or ethnicity, have the ability, as independent individuals and collectively as society, to make, for better or for worse, our social world. It is up to each of us to be the best producers possible of our own developmental course.

EPILOGUE

The professor would like to understand what is not understandable. We ourselves who were there, and who have always asked ourselves the question and will ask it until the end of our lives, we will never understand it, because it cannot be understood.

—*A survivor of the Holocaust*

As I began to write this book, I was reminded of the caution in the above words of an Eastern European Holocaust survivor. The survivor, a physician who had been a prisoner in a Nazi concentration camp, spoke to Robert Lifton as Lifton was gathering material for his 1986 book, *The Nazi Doctors: Medical Killing and the Psychology of Genocide.*

The words, as translated by Lifton,[1] stung and chastened me, as they must have chastened Lifton. As I recounted in the Preface to this book, since my early childhood I also have struggled to understand. My attempts to understand have led me to focus on the pernicious power of the doctrine of biological determinism to distort science and to imbue social policy with heretofore unimaginable, inhumane acts concocted in the name of humanity. Yet I am all too aware that this path to understanding is quite limited. Why, across history, does this doctrine continue to emerge in one transfiguration or another—in White Americans' prejudices toward Native Americans or toward African-Americans, in the Nazi racial hygiene movement, or in contemporary sociobiology? What attracts not just "ordinary" people but also intellectually or politically gifted people to these ideas? Why do so many people remain unaware of or uncaring about the ready co-optation of scientific ideas of biological determinism for politically pernicious, fascist purposes? Cannot society learn, from the folly of its previous involvements with the political co-optation of the doctrine of biological determinism, to take preventive measures against atrocities happening again?

I do not know the answer to these questions, or to the many others that might be asked. Perhaps a message here is that, by making the search for understanding a continual one, we may in fact be helping to ensure that "it doesn't happen again." Perhaps too we must therefore settle for only temporary understanding. If that is so, the only sense I can at this point make out of why Nazi Germany happened is a historical one: We have been warned. We have been given the most draconian of examples, heretofore inconceivable, of the ultimate evils that can be perpetrated in the name of the ideology of biological determinism.

Our response to this warning must be continual vigilance, to paraphrase Richard Lewontin.[2] We must not be afraid to identify other instances of biological determinism and the political co-optation of this view. We must not be intimidated about speaking out about the dangers of such mergers of science and politics—no matter how powerful, prestigious, or eminent the advocates of biological determinism may be.

The survivors of the Holocaust have a duty to bear witness about their past. It is they who most clearly sound the warning. The children that their survival enabled to be brought into life have a duty too. They must look to the future. They must act, as their talents and abilities allow them, to ensure that the phrase "never again" is a constantly renewed reality.

I have found some reassurance that the sense I currently make out of the events of Nazi Germany, a sense gained through my attempts to understand the continuing dangers posed by the political co-optation of the doctrine of biological determinism, is work along a useful path. Ironically, perhaps, it is the words of Lifton that reassure me: "If there is any truth to the psychological and moral judgments we make about the specific and unique characteristics of Nazi mass murder, we are bound to derive from them *principles* that apply more widely—principles that speak to the extraordinary threat and potential for self-annihilation that now haunt human kind."[3]

The threat of the political co-optation of the doctrine of biological determinism is, I believe, one such principle. The enactment of biological determinism into social policies gives us means to make some of our fellow humans less than us. Such prejudice leads inevitably to injustice. All too often it has enabled murder. Any of us, any of our children, can be a target of such politicized science. The elimination of the political and social application of this doctrine may be an act on which the survival of our world truly depends.

NOTES

CHAPTER 1

1. See R. M. Lerner, *Concepts and Theories of Human Development* (New York: Random House, 1986); W. F. Overton, "The Structure of Developmental Theory," in *Annals of Theoretical Psychology*, vol. 7, ed. P. van Geert and L. P. Mos (New York: Plenum, in press); W. F. Overton and H. W. Reese, "Models of Development: Methodological Implications," in *Life-Span Developmental Psychology: Methodological Issues*, ed. J. R. Nesselroade and H. W. Reese (New York: Academic Press, 1973); W. Overton and H. Reese, "Conceptual Prerequisites for an Understanding of Stability-Change and Continuity-Discontinuity," *International Journal of Behavioral Development* 4 (1981): 99–123; H. W. Reese and W. F. Overton, "Models of Development and Theories of Development," in *Life-Span Developmental Psychology: Research and Theory*, ed. L. R. Goulet and P. B. Baltes (New York: Academic Press, 1970).

2. See also Reese and Overton, "Models of Development and Theories of Development"; and Overton and Reese, "Models of Development: Methodological Implications."

3. R. M. Lerner, *Concepts and Theories*; Reese and Overton, "Models of Development."

4. See, e.g., J. Brooks-Gunn and A. C. Petersen, eds., *Girls at Puberty: Biological and Psychosocial Perspectives* (New York: Plenum, 1983); and R. M. Lerner and T. T. Foch, eds., *Biological-Psychosocial Interactions in Early Adolescence* (Hillsdale, N.J.: Erlbaum, 1987).

5. J. M. Tanner, "Physical Growth," in *Carmichael's Manual of Child Psychology*, vol. 1, ed. P. H. Mussen (New York: Wiley, 1970); J. M. Tanner, "Growing Up," *Scientific American* 229 (1973): 34–43.

6. R. M. Lerner and G. B. Spanier, *Adolescent Development: A Life-Span Perspective* (New York: McGraw-Hill, 1980).

7. R. M. Lerner and M. B. Kauffman, "The Concept of Development in Contextualism," *Developmental Review* 5 (1985): 309–33.

8. B. F. Skinner, *Beyond Freedom and Dignity* (New York: Knopf, 1971), 211.

9. See also E. H. Erikson, "Identity and the Life-Cycle," *Psychological Issues* 1 (1959):

18–164; A. L. Gesell, "The Ontogenesis of Infant Behavior," in *Manual of Child Psychology*, ed. L. Carmichael (New York: Wiley, 1946), 295–331.

10. K. W. Schaie, "The Primary Mental Abilities in Adulthood: An Exploration in the Development of Psychometric Intelligence," in *Life-Span Development and Behavior*, vol. 2, ed. P. B. Baltes and O. G. Brim, Jr. (New York: Academic Press, 1979), 67–115.

11. See G. Gottlieb, "Experiential Canalization of Behavioral Development: Results," *Developmental Psychology* 27 (1991): 35–39; R. M. Lerner, D. F. Hultsch, and R. A. Dixon, "Contextualism and the Character of Developmental Psychology in the 1970s," *Annals of the New York Academy of Science* 412 (1983): 101–28; R. Rosnow and M. Georgoudi, eds., *Contextualism and Understanding in Behavioral Research* (New York: Praeger, 1986); T. R. Sarbin, "Contextualism: A World View for Modern Psychology," in *Nebraska Symposium on Motivation, 1976*, ed. J. K. Cole and A. W. Lundfield (Lincoln: University of Nebraska Press, 1977).

12. E. S. Gollin, "Development and Plasticity," in *Developmental Plasticity: Behavioral and Biological Aspects of Variations in Development*, ed. E. S. Gollin (New York: Academic Press, 1981), 231–331; E. Tobach, "Evolutionary Aspects of the Activity of the Organism and Its Development," in *Individuals as Producers of Their Development: A Life-Span Perspective*, ed. R. M. Lerner and N. A. Busch-Rossnagel (New York: Academic Press, 1981), 37–68; E. Tobach and G. Greenberg, "The Significance of T. C. Schneirla's Contribution to the Concept of Levels of Integration," in *Behavioral Evolution and Integrative Levels*, ed. G. Greenberg and E. Tobach (Hillsdale, N.J.: Erlbaum, 1984), 1–7; E. Tobach and T. C. Schneirla, "The Biopsychology of Social Behavior of Animals," in *Biologic Basis of Pediatric Practice*, ed. R. E. Cooke and S. Levin (New York: McGraw-Hill, 1968), 68–82.

13. A. Anastasi, "Heredity, Environment, and the Question 'How,' " *Psychological Review* 65 (1958): 197–208.

14. R. M. Lerner, "Nature, Nurture, and Dynamic Interactionism," *Human Development* 21 (1978): 1–20; R. M. Lerner, "A Dynamic Interactional Concept of Individual and Social Relationship Development," in *Social Exchange in Developing Relationships*, ed. R. Burgess and T. Huston (New York: Academic Press, 1979), 271–305.

15. Tobach, "Evolutionary Aspects."

16. A. Sameroff, "Transactional Models in Early Social Relations," *Human Development* 18 (1975): 65–79.

17. R. M. Lerner, "Nature, Nurture"; R. M. Lerner, "A Dynamic Interactional Concept."

18. S. C. Pepper, *World Hypotheses* (Berkeley and Los Angeles: University of California Press, 1942).

19. Ibid.

20. W. F. Overton, "On the Assumptive Base of the Nature-Nurture Controversy: Additive Versus Interactive Conceptions," *Human Development* 16 (1973): 74–89.

21. See C. M. Super and S. Harkness, "Anthropological Perspectives on Child Development," *New Directions for Child Development* 8 (1980); C. M. Super and S. Harkness, "Figure, Ground, and Gestalt: The Cultural Context of the Active Individual," in *Individuals as Producers of Their Development: A Life-Span Perspective*, ed. R. M. Lerner and N. A. Busch-Rossnagel (New York: Academic Press, 1981), 69–86; C. M. Super and S. Harkness, "The Infant's Niche in Rural Kenya and Metropolitan America," in *Cross-Cultural Research at Issue*, ed. L. L. Adler (New York: Academic Press, 1982), 47–56, for discussions of ethnotheories.

22. S. L. Chorover, *From Genesis to Genocide* (Cambridge, Mass.: MIT Press, 1979); J. Hirsch, "Behavior-Genetic Analysis and Its Biosocial Consequences," *Seminars in Psychiatry* 2 (1970): 89–105; J. Hirsch, "To 'Unfrock the Charlatans,' " *SAGE Race Relations Abstracts* 6 (1981): 1–65; E. Tobach et al., *The Four Horsemen: Racism, Sexism, Militarism, and Social Darwinism* (New York: Behavioral Publications, 1974).

23. S. J. Gould, *The Mismeasure of Man* (New York: Norton, 1981); Hirsch, "To 'Unfrock the Charlatans' "; M. Konner, *The Tangled Wing* (New York: Holt, Rinehart & Winston, 1982); R. C. Lewontin, "On Constraints and Adaptation," *Behavioral and Brain Sciences* 4 (1981): 244–45; E. Staub, *The Roots of Evil: The Origins of Genocide and Other Group Violence* (New York: Cambridge University Press, 1989); Tobach et al., *Four Horsemen*.

24. Chorover, *Genesis to Genocide*; R. C. Lewontin, S. Rose, and L. J. Kamin, *Not in Our Genes: Biology, Ideology, and Human Nature* (New York: Pantheon Books, 1984).

25. See also R. M. Lerner, *On the Nature of Human Plasticity* (New York: Cambridge University Press, 1984); R. M. Lerner, *Concepts and Theories*; O. G. Brim, Jr., and J. Kagan, eds., *Constancy and Change in Human Development* (Cambridge, Mass.: Harvard University Press, 1980).

26. See Konner, *Tangled Wing*; Staub, *Roots of Evil*.

27. For extended discussions, see Chorover, *Genesis to Genocide*; Gould, *Mismeasure of Man*; Lewontin et al., *Not in Our Genes*; B. Müller-Hill, *Murderous Science: Elimination by Scientific Selection of Jews, Gypsies, and Others, Germany 1933–1945*, trans. G. R. Fraser (Oxford: Oxford University Press, 1988); R. N. Proctor, *Racial Hygiene: Medicine Under the Nazis* (Cambridge, Mass.: Harvard University Press, 1988); Tobach et al., *Four Horsemen*.

28. See Chorover, *Genesis to Genocide*; R. J. Lifton, *The Nazi Doctors: Medical Killing and the Psychology of Genocide* (New York: Basic Books, 1986); Lewontin et al., *Not in Our Genes*; Müller-Hill, *Murderous Science*; Proctor, *Racial Hygiene*; Tobach et al., *Four Horsemen*.

29. See R. M. Lerner, *On the Nature of Human Plasticity*.

30. Brim and Kagan, *Constancy and Change*; R. M. Lerner, *On the Nature of Human Plasticity*.

31. T. C. Schneirla, "Interrelationships of the Innate and the Acquired in Instinctive Behavior," in *L'instinct dans le comportement des animaux et de l'homme*, ed. P. P. Grasse (Paris: Mason et Cie, 1956), 387–452; T. C. Schneirla, "The Concept of Development in Comparative Psychology," in *The Concept of Development*, ed. D. B. Harris (Minneapolis: University of Minnesota Press, 1957), 78–108; Tobach, "Evolutionary Aspects"; Tobach and Schneirla, "The Biopsychology of Behaviors in Animals."

32. E. Tobach, Personal communication, November 1990.

33. Staub, *Roots of Evil*.

34. See, e.g., ibid.

35. Konner, *Tangled Wing*, 418.

36. Ibid., 419. See also Staub, *Roots of Evil*, 8.

37. Chorover, *Genesis to Genocide*.

38. A. A. Sicroff, *Les controverses des statuts de "pureté de sang" en Espagna du 15ᵉ au 17ᵉ siècle* (Paris: Didier, 1960).

39. D. Brown, *Bury My Heart at Wounded Knee* (New York: Washington Square Press, 1981), 88.

40. Ibid., 89.

41. Ibid., 166.

42. Lewontin et al., *Not in Our Genes*.

43. Ibid.; Tobach et al., *Four Horsemen*, 97–103.

44. H. Spencer, *The Principles of Sociology*, vols. 1–3 (New York: Appleton, 1910).

45. Tobach et al., *Four Horsemen*, 99.

46. Ibid., 99, 101.

47. Cited in Lewontin et al., *Not in Our Genes*, 26.

48. Ibid., 25–26.

49. D. J. Kevles, *In the Name of Eugenics*, ix.

50. See Lewontin et al., *Not in Our Genes*, 3–15.
51. D. Brown, *The Westerners* (New York: Holt, Rinehart & Winston, 1974).
52. R. G. Hovannisian, ed., *The Armenian Genocide in Perspective* (New Brunswick, N.J.: Transaction Books, 1986); Staub, *Roots of Evil*.
53. M. Billig, *Fascists: A Social Psychological View of the National Front* (New York: Academic Press, 1978).
54. See Lifton, *Nazi Doctors*; Proctor, *Racial Hygiene*; Chorover, *Genesis to Genocide*.
55. See, e.g., Konner, *Tangled Wing*, 419–21.
56. Billig, *Fascists*.
57. Ibid., 6–7.
58. W. L. Shirer, *The Rise and Fall of the Third Reich* (New York: Simon & Schuster, 1960); J. Toland, *Adolf Hitler* (New York: Doubleday, 1976).
59. Shirer, *Rise and Fall of the Third Reich*; Toland, *Hitler*.
60. Billig, *Fascists*, 7.
61. See ibid., 6–7.
62. Ibid., 128.
63. E. O. Wilson, *Sociobiology: The New Synthesis* (Cambridge, Mass.: Harvard University Press, 1975), 4.
64. See, e.g., D. G. Freedman, *Human Sociobiology: A Holistic Approach* (New York: The Free Press, 1979).
65. Ibid., 70, 72.
66. See Proctor, *Racial Hygiene*, 118–30. See also Chapter 3 in the present book.

CHAPTER 2

1. Proctor, *Racial Hygiene*, 10. Hess is quoted in Lifton, *Nazi Doctors*, 31.
2. Ibid., 63, 352.
3. Ibid., 7.
4. Ibid., 64; Lifton, *Nazi Doctors*, 31.
5. A. Hitler, *Mein Kampf*, trans. R. Manheim (Boston: Houghton Mifflin, 1925, 1927/ 1943).
6. Ibid., 132, 135.
7. Lifton, *Nazi Doctors*, 14.
8. D. Gasman, *The Scientific Origins of Natural Socialism: Social Darwinism in Ernst Haeckel and the German Monist League* (New York: Elsevier, 1971); T. J. Kalikow, "Konrad Lorenz's Ethological Theory: Explanation and Ideology, 1938–1943," *Journal of the History of Biology* 16 (1983): 39–73.
9. There is some controversy about the precise extent of Haeckel's influence on Nazi ideology. For example, compare Gasman's *Scientific Origins*, Kalikow's "Lorenz's Ethological Theory," and G. J. Stein's "The Biological Bases of Ethnocentrism, Racism, and Nationalism in National Socialism," in *The Sociobiology of Ethnocentrism*, ed. V. Reynolds et al. (London: Croom Helm, 1987), 251–67, with R. J. Richards's *Darwin and the Emergence of Evolutionary Theories of Mind and Behavior* (Chicago: University of Chicago Press, 1987). However, all reviewers acknowledge at least some important contributions both of Haeckel and of other members of his Monist League (see note 12, below). The issue of the relationship between Haeckelian / Monist League ideas and Nazi ideology will be discussed in greater detail in Chapter 3, in the context of a presentation of the views of Konrad Lorenz.
10. Stein, "Biological Bases," 259.

11. Ibid.

12. Stein, "Biological Bases." Members of Haeckel's Monist League included H. Ziegler, a biologist; A. Kalthoff, a Protestant theologian; H. Schmidt, a professor at Jena University and ultimately Haeckel's biographer; W. Boelsche, a literary critic and novelist; B. Wille, also a literary critic and novelist; R. H. France, a biologist and the first editor of the monthly publication *Der Monismus;* J. Unold, a physician and Social Darwinist political writer; A. Forel, a geneticist; W. Schallmayer, a founding figure in the German eugenics movement; L. Gurlitt, a co-founder of the *Wandervogel;* L. Plate, a biologist and successor to Haeckel's chair at Jena; and W. Ostwald, a Nobel Prize winner in chemistry (Richards, "The Natural Selection Model"; Stein, "Biological Bases").

13. Stein, "Biological Bases," 261–62.

14. Ibid., 262, quoting E. Haeckel, *The History of Creation; or, The Development of the Earth and Its Inhabitants by the Action of Natural Causes* (New York: Appleton, 1876), 434, 332, 310, and 390, respectively.

15. See Haeckel, *History of Creation,* 170; E. Haeckel, *The Wonders of Life* (New York: Harper, 1905), 116, 118–19; and Stein, "Biological Bases," 264.

16. Stein, "Biological Bases."

17. See, respectively, Haeckel, *Wonders of Life,* 116, 118–19; and Haeckel, *History of Creation,* 172–73. See Stein, "Biological Bases," for a fuller presentation of Haeckel's views about the use of killing in implementing a negative eugenics program.

18. Stein, "Biological Bases," 265.

19. J. P. Rushton, "Race Differences in Behavior: A Review and Evolutionary Analysis," *Personality and Individual Differences* 9 (1988): 1009–24.

20. Lifton, *Nazi Doctors,* 441.

21. Ibid.

22. Ibid., 478.

23. See L. S. Dawidowicz, *The War Against the Jews, 1933–1945* (New York: Holt, Rinehart & Winston, 1975), 31.

24. Lifton, *Nazi Doctors,* 478.

25. Dawidowicz, *War Against the Jews,* 31.

26. Cited in ibid., 32.

27. See Proctor, *Racial Hygiene;* Chorover, *Genesis to Genocide.*

28. Cited in Lifton, *Nazi Doctors,* 24.

29. Chorover, *Genesis to Genocide,* 94.

30. Proctor, *Racial Hygiene,* 15.

31. Ibid.

32. Ibid.

33. Ibid.

34. Ibid., 15–16.

35. Chorover, *Genesis to Genocide,* 95.

36. Lifton, *Nazi Doctors.*

37. Chorover, *Genesis to Genocide,* 97.

38. Chorover, *Genesis to Genocide;* Lifton, *Nazi Doctors;* Müller-Hill, *Murderous Science;* Proctor, *Racial Hygiene.*

39. See Kevles, *In the Name of Eugenics,* for a review.

40. T. S. Kuhn, *The Structure of Scientific Revolutions,* 2nd ed. (Chicago: University of Chicago Press, 1970); Proctor, *Racial Hygiene.*

41. Proctor, *Racial Hygiene.*

42. See ibid., 284–85.

43. Ibid., 49.

44. Quoted in ibid., 50.

45. F. Lenz, "Alfred Ploetz zum 70. Geburtstag am 22 August," *Archiv für Rassen- und Gesellschaftsbiologie* 24 (1930): xiv, translated by Proctor, *Racial Hygiene*, 49.

46. Ibid., 49.

47. See ibid.

48. Lifton, *Nazi Doctors*, 48.

49. Ibid., 46.

50. Proctor, *Racial Hygiene*, 203.

51. Chorover, *Genesis to Genocide*, 102.

52. Proctor, *Racial Hygiene*, 30.

53. Translated in ibid., 286.

54. Ibid., 286.

55. See ibid., 131–36.

56. Hitler, *Mein Kampf*, 403–4.

57. Cited in M. Gilbert, *The Holocaust: A History of the Jews of Europe During the Second World War* (New York: Holt, Rinehart & Winston, 1985), 76.

58. Cited in Dawidowicz, *War Against the Jews*, 64.

59. Translated by Gilbert, *The Holocaust*, 245–46.

60. Ibid., 615–16.

61. Hitler, *Mein Kampf*, 435.

62. See Dawidowicz, *War Against the Jews*; Lifton, *Nazi Doctors*.

63. Lifton, *Nazi Doctors*, 16.

64. Y. Bauer, "Genocide: Was It the Nazis' Original Plan?" *Annals of the American Academy of Political and Social Science* 450 (1980): 37.

65. Translated by George R. Fraser in Müller-Hill, *Murderous Science*, 12.

66. Bauer, "Genocide," 38.

67. Ibid., 44.

68. Translated in S. Friedlander, "From Anti-Semitism to Extermination," in *Unanswered Questions: Nazi Germany and the Genocide of the Jews*, ed. F. Furet (New York: Schocken Books, 1989), 7.

69. E. Nolte, *Three Faces of Fascism: Action Française, Italian Fascism, National Socialism* (New York: Holt, Rinehart & Winston, 1965), 400.

70. R. Hilberg, *The Destruction of the European Jews* (Chicago: Quadrangle Books, 1961).

71. Friedlander, "From Anti-Semitism to Extermination," 328.

72. Ibid.

73. Ibid.

74. Ibid., 327.

75. Dawidowicz, *War Against the Jews*, 58.

76. Ibid., 59.

77. Ibid., 60–61.

78. Ibid., 63.

79. Proctor, *Racial Hygiene*, 103.

80. K. A. Schleunes, "Retracing the Twisted Road," in *Unanswered Questions: Nazi Germany and the Genocide of the Jews*, ed. F. Furet (New York: Schocken Books, 1989), 62.

81. Dawidowicz, *War Against the Jews*.

82. Hitler, *Mein Kampf*, 640.

83. Bauer, "Genocide"; A. Bullock, *Hitler: A Study in Tyranny*, rev. ed. (New York: Harper & Row, 1962).

84. Hitler, *Mein Kampf*, 324, 214, and 65, respectively.

85. Ibid., 324.
86. Ibid., 213. Hitler cites Frederick the Great and Richard Wagner, along with Martin Luther, as "truly great statesmen."
87. Translated by Dawidowicz, *War Against the Jews*, 23.
88. Translated by Lifton, *Nazi Doctors*, 479.
89. Toland, *Adolf Hitler*.
90. See S. Volkov, "The Written Matter and the Spoken Word," in *Unanswered Questions: Nazi Germany and the Genocide of the Jews*, ed. F. Furet (New York: Schocken Books, 1989), 33–53.
91. Ibid., 52 (emphasis in the original).
92. C. R. Browning, "The Decision Concerning the Final Solution," in *Unanswered Questions: Nazi Germany and the Genocide of the Jews*, ed. F. Furet (New York: Schocken Books, 1989), 100.
93. R. Wolfe, "Putative Threat to National Security as a Nuremberg Defense for Genocide," *Annals of the American Academy of Political and Social Science* 450 (1980): 46.
94. Ibid., 54.
95. Ohlendorf quotations translated by Wolfe, "Putative Threat," 64, 66.
96. Translated by Shirer, *Rise and Fall of the Third Reich*, 959.
97. Proctor, *Racial Hygiene*, 195–96.
98. Lifton, *Nazi Doctors*, 448.
99. Proctor, *Racial Hygiene*, 204–5.
100. Translated in ibid., 37–38.
101. Ibid., 204–5.
102. Ibid., 176, 379.
103. Dawidowicz, *War Against the Jews*.
104. Translated by Lifton, *Nazi Doctors*, 157.
105. Ibid.; Proctor, *Racial Hygiene*.
106. Lifton, *Nazi Doctors*; Müller-Hill, *Murderous Science*; Proctor, *Racial Hygiene*.
107. Lifton, *Nazi Doctors*.
108. Translated in ibid., 30.
109. Translated by Chorover, *Genesis to Genocide*, 100.
110. Ibid.
111. Translated by Lifton, *Nazi Doctors*, 30.
112. Ibid., 412.
113. Ibid., 413.
114. Translated by Gilbert, *The Holocaust*, 457.
115. Translated by Lifton, *Nazi Doctors*, 96.
116. Proctor, *Racial Hygiene*, 50.
117. Lifton, *Nazi Doctors*, 431.
118. Translated in ibid., 16.
119. Hitler, *Mein Kampf*, 397–98.
120. Lifton, *Nazi Doctors*; Müller-Hill, *Murderous Science*; Proctor, *Racial Hygiene*; Shirer, *Rise and Fall of the Third Reich*.
121. See, e.g., Chorover, *Genesis to Genocide*; Kalikow, "Lorenz's Ethological Theory."

CHAPTER 3

1. M. Weinreich, *Hitler's Professors: The Part of Scholarship in Germany's Crimes Against the Jewish People* (New York: Yiddish Scientific Institute, 1946).

2. Proctor, *Racial Hygiene*, 299.

3. See Lifton, *Nazi Doctors*; Müller-Hill, *Murderous Science*; and Proctor, *Racial Hygiene*, for documentation and examples of what has happened to participants in the Nazi agenda.

4. Richards, *Darwin and the Emergence of Evolutionary Theories*, 528.

5. See, e.g., K. Lorenz, "A Consideration of Methods of Identification of Species-Specific Instinctive Patterns in Birds," in *Studies in Animal and Human Behavior*, ed. and trans. R. Martin (Cambridge, Mass.: Harvard University Press, 1932/1970), 57–100; K. Lorenz, "Über die Bildung des Instinktbegriffes," *Die Naturwissenschaften* 25 (1937): 189–300, 307–8, 324–33. Lorenz's five criteria are in the latter. See also K. Lorenz, "Psychologie und Stammesgeschichte," in *Die Evolution der Organismen*, 2nd ed., ed. G. Heberer (Jena: G. Fischer, 1954), 131–77.

6. Richards, *Darwin and the Emergence of Evolutionary Theories*, 530.

7. Lorenz, "Über die Bildung."

8. Richards, *Darwin and the Emergence of Evolutionary Theories*.

9. D. S. Lehrman, "Semantic and Conceptual Issues in the Nature-Nurture Problem, in *Development and Evolution of Behavior: Essays in Memory of T. C. Schneirla*, ed. L. R. Aronson et al. (San Francisco: Freeman, 1970).

10. K. Lorenz, *Evolution and Modification of Behavior* (Chicago: University of Chicago Press, 1965), 103.

11. Lehrman, "Semantic and Conceptual Issues," 24.

12. Richards, *Darwin and the Emergence of Evolutionary Theories*, 530–31; Lorenz, *Evolution and Modification of Behavior*, 51.

13. Richards, *Darwin and the Emergence of Evolutionary Theories*, 531.

14. E.g., K. Lorenz, "Über den Begriff der Instinkthandlung," *Folia Biotheoretica* 2 (1937): 17–50.

15. Lorenz, *Evolution and Modification of Behavior*; K. Lorenz, *On Aggression* (New York: Harcourt, Brace & World, 1966).

16. E.g., D. O. Hebb, *The Organization of Behavior* (New York: Wiley, 1949); D. S. Lehrman, "A Critique of Konrad Lorenz's Theory of Instinctive Behavior," *Quarterly Review of Biology* 28 (1953): 337–63; Lehrman, "Semantic and Conceptual Issues"; Schneirla, "Interrelationships of the Innate"; Schneirla, "Concept of Development."

17. E.g., Lehrman, "Semantic and Conceptual Issues."

18. E.g., Hebb, *Organization of Behavior*.

19. E.g., Schneirla, "Interrelationships of the Innate"; Schneirla, "Concept of Development."

20. Lorenz, *Evolution and Modification of Behavior*, 1.

21. Ibid., 79.

22. Ibid.

23. Compare with Richards, *Darwin and the Emergence of Evolutionary Theories*, 530.

24. Lorenz, *Evolution and Modification of Behavior*, 13.

25. Kalikow, *Lorenz's Ethological Theory*, 39.

26. Ibid.

27. On Lorenz's hypothesis here, see Richards, *Darwin and the Emergence of Evolutionary Theories*, 532.

28. Lorenz, quoted in V. Cox, "A Prize for the Goose Father," *Human Behavior* 3 (1974): 20. Cox's article is a report on a 1974 interview with Lorenz.

29. See T. J. Kalikow, "Konrad Lorenz's 'Brown Past': A Reply to Alec Nisbett," *Journal of the History of the Behavioral Sciences* 14 (1978): 173–79; Kalikow, "Lorenz's Ethological Theory."

30. Kalikow, "Lorenz's 'Brown Past' "; and, e.g., Chorover's *Genesis to Genocide* and L. Eisenberg's "On the Human Nature of Human Nature," *Science* 176 (1972): 123–28.

31. Kalikow, "Lorenz's 'Brown Past' "; Kalikow, "Lorenz's Ethological Theory"; and, e.g., A. Nisbett, *Konrad Lorenz: A Biography* (New York: Harcourt Brace Jovanovich, 1977); Chorover, *Genesis to Genocide*; Lewontin et al., *Not in Our Genes*; Lifton, *Nazi Doctors*; Proctor, *Racial Hygiene*.

32. Kalikow, "Lorenz's Ethological Theory," 56.

33. See ibid., 58–61.

34. Translations from Lorenz's 1938 paper are in ibid., 58–61.

35. Ibid., 61.

36. Translated by Kalikow, "Lorenz's 'Brown Past,' " 174–75; from Lorenz's 1938 paper, "Über Ausfallserscheinungen," 146–47.

37. Translated by Kalikow, "Lorenz's Ethological Theory," 63.

38. Ibid., 62.

39. For the reviewers, see Kalikow, "Lorenz's 'Brown Past' "; Kalikow, "Lorenz's Ethological Theory"; Nisbett, *Lorenz: A Biography*; Cox, "A Prize"; Chorover, *Genesis to Genocide*. For Lorenz, see quotations in Cox, "A Prize," and in R. I. Evans, "Interview with Konrad Lorenz," *Psychology Today* 8 (1974): 26ff.

40. See Kalikow, "Lorenz's Ethological Theory."

41. Translated by Kalikow, "Lorenz's 'Brown Past,' " 176 (emphasis in the original).

42. Translated by Eisenberg, "On Human Nature," 124.

43. Translated by Chorover, *Genesis to Genocide*, 105.

44. K. Lorenz, "Durch Domestikation verursachte Störungen arteigenen Verhaltens," *Zeitschrift für angewandte Psychologie und Charakterkunde* 59 (1940): 2–81; translated by George R. Fraser in Müller-Hill, *Murderous Science*, 14.

45. Quoted in Cox, "A Prize," 20.

46. Translated by Kalikow, "Lorenz's Ethological Theory," 68 (emphasis in the original).

47. Ibid.

48. Translated by Kalikow, "Lorenz's 'Brown Past,' " 177–78.

49. See ibid.; Kalikow, "Lorenz's Ethological Theory."

50. Chicago: Chicago University Press, 1987.

51. Ibid.; see Kalikow, "Lorenz's Ethological Theory."

52. Gasman, *The Scientific Origins*.

53. Kalikow, "Lorenz's Ethological Theory," 46–48.

54. Richards, *Darwin and the Emergence of Evolutionary Theories*. See Kalikow, "Lorenz's 'Brown Past' "; Kalikow, "Lorenz's Ethological Theory."

55. Richards, *Darwin and the Emergence of Evolutionary Theories*, 533.

56. Ibid. See Kalikow, "Lorenz's Ethological Theory."

57. Richards, *Darwin and the Emergence of Evolutionary Theories*. See Kalikow, "Lorenz's Ethological Theory."

58. Manheim, in Hitler, *Mein Kampf*, xi–xii.

59. Richards, *Darwin and the Emergence of Evolutionary Theories*, 533; Kalikow, "Lorenz's Ethological Theory," 62.

60. Müller-Hill, *Murderous Science*.

61. K. Binding and A. Hoche, *Die Freigabe der Vernichtung lebensunwerten Lebens* (The Sanctioning of the Destruction of Lives Unworthy to Be Lived) (Leipzig: F. Meiner, 1920). See also Proctor, *Racial Hygiene*.

62. Hitler, *Mein Kampf*, 131–32.

63. Ibid., 255.

64. Lifton, *Nazi Doctors*, 45–79.

65. Hitler, *Mein Kampf*; Lorenz, "Durch Domestikation"; Lorenz, *On Aggression*.

66. Hitler, *Mein Kampf*, 135.

67. H. Ziegler, *Die Naturwissenschaft und die Socialdemokratische Theorie* (Stuttgart: Enke, 1893), 168–69.

68. Lorenz, *On Aggression*, 251, 48.

69. Richards, *Darwin and the Emergence of Evolutionary Theories*, 533.

70. See ibid.; Ziegler, *Die Naturwissenschaft*.

71. Lorenz, *On Aggression*, 278.

72. Kalikow, "Lorenz's Ethological Theory."

73. Richards, *Darwin and the Emergence of Evolutionary Theories*, 533.

74. Ibid., 534.

75. Ibid., 531; Kalikow, "Lorenz's Ethological Theory." See Lorenz, "Über den Begriff."

76. Kalikow, "Lorenz's Ethological Theory"; Nisbett, *A Biography*.

77. Richards, *Darwin and the Emergence of Evolutionary Theories*, 535.

78. Ibid.

79. I am grateful to Professor Alexander von Eye for pointing out this proverb to me in this context.

80. Richards, *Darwin and the Emergence of Evolutionary Theories*, 535–36.

81. Kalikow, "Lorenz's 'Brown Past' "; Kalikow, "Lorenz's Ethological Theory"; Müller-Hill, *Murderous Science*.

82. Richards, *Darwin and the Emergence of Evolutionary Theories*, 536.

83. K. Lorenz, "Systematik und Entwicklungsgedanke im Unterricht," *Der Biologe 9* (1940): 24–36; translated by George R. Fraser in Müller-Hill, *Murderous Science*, 184.

84. Kalikow, "Lorenz's 'Brown Past' "; Kalikow, "Lorenz's Ethological Theory."

85. See, e.g., K. Lorenz, "Kants Lehre vom Apriorischen im Lichte gegenwärtiger Biologie," *Blätter für deutsche Philosophie* 15 (1941): 94–125; and Lorenz, *Eight Deadly Sins*.

86. Quoted in Cox, "A Prize," 20.

87. Ibid.

88. Ibid.

89. K. Lorenz, "Psychologie und Stammesgeschichte," in *Die Evolution der Organismen*, 2nd ed., ed. G. Heberer (Jena: G. Fischer, 1954), 131–77, translated by Kalikow, "Lorenz's Ethological Theory," 70.

90. K. Lorenz, "Ganzheit und Teil in der tierischen und menschlichen Gemeinschaft," *Studium Generale 9* (1950): 184; translated by T. J. Kalikow, "Book Review of *Civilized Man's Eight Deadly Sins*," *Philosophy of the Social Sciences* 8 (1978): 100.

91. See, e.g., Lorenz, *On Aggression*, 270–74.

92. Ibid., 270.

93. Ibid., 270.

94. Ibid., 270–71.

95. Ibid., 271.

96. Ibid., 278, 281.

97. Ibid., 272.

98. For a description of Lorenz's views about the reflex-like character of militant enthusiasm, see J.M.G. van der Dennen, "Ethnocentrism and In-Group/Out-Group Differentiation: A Review and Interpretation of the Literature," in *The Sociobiology of Ethnocentrism*, ed. V. Reynolds et al. (London: Croom Helm, 1987), 11.

99. Lorenz, *On Aggression*, 272, 273.

100. See Dawidowicz, *War Against the Jews*.

101. T. C. Schneirla, "Instinct and Aggression," *Natural History 75* (1966): 16ff.

102. Tobach et al., *Four Horsemen*.

103. Lorenz, *Evolution and Modification*; Richards, *Darwin and the Emergence of Evolutionary Theories*.

104. Lorenz, *On Aggression*; K. Lorenz, "Letter: Lorenz Clarifies Ideas," *Human Behavior*, September 1974, p. 6.

105. Lorenz, *Eight Deadly Sins*.

106. See Lorenz, "Durch Domestikation."

107. K. Lorenz, "Konrad Lorenz Responds to Donald Campbell," in *Konrad Lorenz: The Man and His Ideas*, ed. R. I. Evans (New York: Harcourt Brace Jovanovich, 1975), 126–28.

108. See Lenz, "Alfred Ploetz"; Proctor, *Racial Hygiene*.

109. Cox, "A Prize," 19. See Lorenz, "Durch Domestikation."

110. Lorenz, quoted in Cox, "A Prize," 19.

111. As translated by Kalikow, "Lorenz's Ethological Theory," 66, *and* by Lorenz, *Eight Deadly Sins*, 6.

112. Lorenz, *Eight Deadly Sins*.

113. Wilson, *The New Synthesis*; E. O. Wilson, "Human Decency Is Animal," *New York Times Magazine*, October 12, 1975; E. O. Wilson, "For Sociobiology," *New York Review of Books*, December 11, 1975.

114. E.g., G. W. Barlow, "The Development of Sociobiology: A Biologist's Perspective," in *Sociobiology: Beyond Nature/Nurture?* ed. G. W. Barlow and J. Silverberg (Boulder, Colo.: Westview Press, 1980), 3–24; A. L. Caplan, ed., *The Sociobiology Debate* (New York: Harper & Row, 1978); A. L. Caplan, "A Critical Examination of Current Sociobiological Theory: Adequacy and Implications," in *Sociobiology: Beyond Nature/Nurture?* ed. G. W. Barlow and J. Silverberg (Boulder, Colo.: Westview Press, 1980), 97–121; R. Dawkins, *The Selfish Gene* (New York: Oxford University Press, 1976); Freedman, *Human Sociobiology*; Konner, *Tangled Wing*.

115. Konner, *Tangled Wing*, 124. See K. Lorenz, "Der Kumpan in der Umweld des Vogels," *Journal für Ornithologie* 83 (1935): 137–215, 289–413.

CHAPTER 4

1. Wilson, *Sociobiology: The New Synthesis* (Cambridge, Mass.: Harvard University Press, 1975).

2. R.I.M. Dunbar, "Sociobiological Explanations and the Evolution of Ethnocentrism," in *The Sociobiology of Ethnocentrism*, ed. V. Reynolds et al. (London: Croom Helm, 1987), 51. In addition, scholars from outside of sociobiology have criticized this field. For instance, see Lewontin et al., *Not in Our Genes*, and P. Kitcher, *Vaulting Ambition* (Cambridge, Mass., MIT Press, 1985).

3. Ibid.

4. C. J. Lumsden and E. O. Wilson, *Genes, Mind, and Culture* (Cambridge, Mass.: Harvard University Press, 1981).

5. See, e.g., Wilson, *Sociobiology: The New Synthesis*; Dawkins, *Selfish Gene*; D. P. Barash, *Sociobiology and Behavior* (New York: Elsevier, 1977); and Freedman, *Human Sociobiology*.

6. Konner, *Tangled Wing*, 265.

7. Dawkins, *Selfish Gene*, ix.

8. Wilson, *Sociobiology: The New Synthesis*, 3.

9. Dawkins, *Selfish Gene*, 21.

10. See, e.g., ibid., ix.

11. For the term "blind," see ibid., ix.

12. Ibid., 71.

13. Konner, *Tangled Wing*, 203.

14. Ibid., xviii.

15. H. Flohr, "Biological Bases of Social Prejudices," in *The Sociobiology of Ethnocentrism,* ed. V. Reynolds et al. (London: Croom Helm, 1987), 195.

16. E. O. Wilson, *On Human Nature* (Cambridge, Mass.: Harvard University Press, 1978). On Dawkins, see Flohr, "Social Prejudices," 199.

17. Wilson, "Academic Vigilantism," 70.

18. Konner, *Tangled Wing*, xviii.

19. Freedman, *Human Sociobiology*, 2.

20. R. Trivers, in *Sociobiology: Doing What Comes Naturally,* a film distributed by Document Associates, Inc., 880 Third Avenue, New York, N.Y., 1974.

21. Freedman, *Human Sociobiology*, 12.

22. P. L. van den Berghe and D. P. Barash, "Inclusive Fitness and Human Family Structure," *American Anthropologist* 79 (1977): 813.

23. Ibid., 813, 814, 815.

24. Ibid., 813.

25. Barash, *Sociobiology and Behavior*, xv.

26. Van den Berghe and Barash, "Inclusive Fitness," 814–15.

27. Dawkins, *Selfish Gene*, 152, 153, 157–58.

28. Van den Berghe and Barash, "Inclusive Fitness," 814.

29. Freedman, *Human Sociobiology*, 14, 19, 20–21; for Kinsey's findings, see A. C. Kinsey et al., *Sexual Behavior in the Human Female* (Philadelphia: Saunders, 1953).

30. Dawkins, *Selfish Gene*, 136.

31. Dawkins, *Selfish Gene; * Freedman, *Human Sociobiology; * van den Berghe and Barash, "Inclusive Fitness." See also S. K. Kachigan, *The Sexual Matrix* (New York: Radius Press, 1990).

32. Kachigan, *Sexual Matrix*, 55, 105 (emphasis in the original).

33. Wilson, *Sociobiology: The New Synthesis; * Wilson, "Human Decency"; S. Freud, *The Ego and the Id* (London: Hogarth, 1923); E. H. Erikson, *Identity, Youth, and Crisis* (New York: Norton, 1968).

34. See, e.g., Lumsden and Wilson, *Genes, Mind, and Culture.*

35. Dawkins, *Selfish Gene*, 65–66 and 3, respectively.

36. Wilson, *Sociobiology: The New Synthesis*, 4.

37. E. O. Wilson, "A Consideration of the Genetic Foundation of Human Social Behavior," in *Sociobiology: Beyond Nature/Nurture,* ed. G. W. Barlow and J. Silverberg (Boulder, Colo.: Westview Press, 1980), 296. Critics were, e.g., E. Allen et al., "Against 'Sociobiology,' " *New York Review of Books* 182 (1975): 184–86. See also Sociobiology Study Group of Science for the People, "Sociobiology: Another Biological Determinism," *BioScience* 26 (March 1976).

38. Wilson, "Consideration of the Genetic Foundation," 296.

39. Freedman, *Human Sociobiology*, 13. See A. J. Bateman, "Intrasexual Selection of Drosophila," *Heredity* 2 (1948): 349–68; C. B. Koford, "Group Relations in an Island Colony of Rhesus Monkeys," in *Prime Social Behavior,* ed. C. H. Southwick (Princeton: Van Nostrand, 1963), 136–52; and N. A. Chagnon, *Yanomamo: The Fierce People* (New York: Holt, Rinehart & Winston, 1972).

40. Freedman, *Human Sociobiology*, 14.

41. B. Grzimek, ed., *Animal Life Encyclopedia,* vol. 7 (New York: Van Nostrand Reinhold, 1977), 270.

42. Freedman, *Human Sociobiology*, 14.

43. See Wilson, *Sociobiology: The New Synthesis*; Barash, *Sociobiology and Behavior*.

44. See J. W. Atz, "The Application of the Idea of Homology to Behavior," in *Development and Evolution of Behavior: Essays in Memory of T. C. Schneirla*, ed. L. R. Aronson et al. (San Francisco: Freeman, 1970), 53–74; S. J. Gould, "Sociobiology and the Theory of Natural Selection," in *Sociobiology: Beyond Nature/Nurture*, ed. G. W. Barlow and J. Silverberg (Boulder, Colo.: Westview Press, 1980), 257–69.

45. See M. E. Bitterman, "Phyletic Differences in Learning," *American Psychologist* 20 (1965): 396–410; M. E. Bitterman, "The Comparative Analysis of Learning," *Science* 18 (1975): 699–709.

46. See J. Piaget, *The Psychology of Intelligence* (New York: Harcourt, Brace, 1950); J. Piaget, "Piaget's Theory," in *Carmichael's Manual of Child Psychology*, vol. 1 (New York: Wiley, 1970), 703–32.

47. See Atz, "Homology to Behavior"; Schneirla, "Concept of Development."

48. Dunbar, "Sociobiological Explanations," 53.

49. See S. Brownmiller, *Against Our Will: Men, Women, and Rape* (New York: Simon & Schuster, 1975); S. R. Sunday and E. Tobach, eds., *Violence Against Women: A Critique of the Sociobiology of Rape* (New York: Gordian Press, 1985).

50. E.g., compare Wilson's *Sociobiology: The New Synthesis* with Dunbar's "Sociobiological Explanations" on whether sociobiology will replace the social and behavioral sciences.

51. Wilson, "Consideration of the Genetic Foundation," 297.

52. Ibid., 299.

53. See, e.g., A. R. Jensen, "How Much Can We Boost IQ and Scholastic Achievement?" *Harvard Educational Review* 39 (1969): 1–123.

54. D. O. Hebb, "A Return to Jensen and His Social Critics," *American Psychologist* 25 (1970): 568.

55. Lehrman, "Semantic and Conceptual Issues."

56. S. Scarr-Salapatek, "Unknowns in the IQ Equation," *Science* 174 (1971): 1128.

57. See, e.g., Hirsch, "Behavior-Genetic Analysis"; J. Hirsch, "Review of Wilson's *Sociobiology*," *Animal Behavior* 24 (1976): 707–9; J. Hirsch, "Correlation, Causation, and Careerism," *European Bulletin of Cognitive Psychology* 10 (1990): 647–52; J. Hirsch, "A Nemesis for Heritability Estimation," *Behavioral and Brain Sciences* 13 (1990): 137–38; J. Hirsch, "Obfuscation of Interaction Through Incantation" (Manuscript, University of Illinois, Urbana, 1990); J. Hirsch, T. R. McGuire, and A. Vetta, "Concepts of Behavior Genetics and Misapplications of Humans," in *The Evolution of Human Social Behavior*, ed. J. S. Lockard (New York: Elsevier, 1980), 215–38; R. M. Lerner, *Concepts and Theories*; T. R. McGuire and J. Hirsch, "General Intelligence (*g*) and Heritability (*H²*, *h²*)," in *The Structuring of Experience*, ed. I. C. Uzgiris and F. Weizmann (New York: Plenum, 1977), 25–72; D. Walsten, "Insensitivity of the Analysis of Variance to Heredity-Environment Interaction," *Behavioral and Brain Sciences* 13 (1990): 109–20.

58. McGuire and Hirsch, "General Intelligence," 46.

59. Ibid.

60. Walsten, "Insensitivity of the Analysis of Variance."

61. D. Bullock, "Methodological Heterogeneity and the Anachronistic Status of ANOVA in Psychology," *Behavioral and Brain Sciences* 13 (1990): 122–23; M. W. Feldman and R. C. Lewontin, "The Heritability Hang-Up," *Science* 190 (1975): 1163–68; McGuire and Hirsch, "General Intelligence"; Walsten, "Insensitivity of the Analysis of Variance."

62. Walsten, "Insensitivity of the Analysis of Variance."

63. Ibid.

64. Hirsch et al., "Concepts of Behavior," 236.

65. Hirsch et al., "Concepts of Behavior."

66. E. E. Philipp, "Discussion," in *Law and Ethics of AID and Embryo Transfer,* Ciba Foundation Symposium 17, new series (Amsterdam: Elsevier North Holland, 1973), 66. See also Hirsch, "Correlation, Causation, and Careerism"; and Hirsch et al., "Concepts of Behavior."

67. O. Kempthorne, "How Does One Apply Statistical Analysis to Our Understanding of the Development of Human Relationships?" *Behavioral and Brain Sciences* 13 (1990): 139.

68. Feldman and Lewontin, "Heritability Hang-Up," 1168.

69. Lewontin et al., *Not in Our Genes.*

70. See Wilson, *Sociobiology: The New Synthesis.*

71. Hirsch, "Review," 707.

72. D. S. Falconer, *Introduction to Quantitative Genetics* (New York: Ronald Press, 1960), 167.

73. Hirsch et al., "Concepts of Behavior," 215.

74. Gould, "Sociobiology and the Theory of Natural Selection," 258.

75. L. von Bertalanffy, "Chance or Law?" in *Beyond Reductionism,* ed. A. Koestler (London: Hitchinson, 1969), 11.

76. Dunbar, "Sociobiological Explanations," 50.

77. Konner, *Tangled Wing,* 18.

78. See S. Gould and E. Vrba, "Exaptation: A Missing Term in the Science of Form," *Paleobiology* 8 (1982): 4–15.

79. G. C. Williams, *Adaptation and Natural Selection* (Princeton: Princeton University Press, 1966), 6.

80. See W. Bock, "The Use of Adaptive Characters in Avian Classification," *Proceedings of the XIV International Ornithology Congress, 1967;* W. Bock, "Synthetic Explanation of Macroevolutionary Change—A Reductionistic Approach," *Bulletin of the Carnegie Museum of Natural History* 13 (1979): 20–69; W. J. Bock, "The Definition and Recognition of Biological Adaptation," *American Zoologist* 20 (1980): 217–27.

81. Bock, "A Synthetic Explanation," 39.

82. Gould and Vrba, "Explanation."

83. C. Darwin, *On the Origin of Species by Means of Natural Selection; or, The Preservation of Favored Races in the Struggle for Life* (London: John Murray, 1859), 197.

84. See Williams, *Adaptation and Natural Selection.*

85. Gould and Vrba, "Exaptation."

86. Ibid., 6. See Williams, *Adaptation and Natural Selection.*

87. See S. J. Gould and R. C. Lewontin, "The Spandrels of San Marco and the Panglossian Paradigm: A Critique of the Adaptationist Programme," in *The Evolution of Adaptation by Natural Selection,* ed. J. Maynard Smith and R. Holliday (London: Royal Society of London, 1979), 581–98; Konner, *Tangled Wing,* 18.

88. Lewontin, "On Constraints and Adaptation."

89. Ibid., 245.

90. Gould and Vrba, "Exaptation," 7.

91. Preadaptation is an invalid concept both in my view (see, e.g., R. M. Lerner, *Human Plasticity*) and in that of Gould and Vrba, "Exaptation." The concept involves invoking teleology to explain how a future and thus empirically nonexistent "state of affairs" shapes the current status of a feature of the organism. In other words, the future is used to explain the present. In being nonempirical, the concept of preadaptation is not scientifically useful.

92. Gould and Vrba, "Exaptation," 12, 13.

CHAPTER 5

1. Müller-Hill, *Murderous Science;* Proctor, *Racial Hygiene.*

2. See Billig, *Fascists;* Gould, *Mismeasure of Man;* Hovannisian, *Armenian Genocide;*

Hirsch, "Charlatans"; Kevles, *Name of Eugenics*; Konner, *Tangled Wing*; Lewontin et al., *Not in Our Genes*; Staub, *Roots of Evil*; and Sunday and Tobach, *Violence Against Women*, for other examples.

3. See, e.g., Wilson, "For Sociobiology"; Wilson, "Academic Vigilantism."

4. See P. L. van den Berghe, "Why Most Sociologists Don't (and Won't) Think Evolutionarily," *Sociological Forum* 5 (1990): 179; Wilson, "Academic Vigilantism."

5. See J. P. Rushton, "An Evolutionary Theory of Health, Longevity, and Personality: Sociobiology and r/K Reproductive Strategies," *Psychological Reports* 60 (1987): 539–49; Rushton, "Race Differences"; J. P. Rushton, "Do r/K Reproductive Strategies Apply to Human Differences?" *Social Biology* 35 (1988): 337–40.

6. Quoted in Canada's national newspaper, *The Globe and Mail*, February 4, 1989.

7. E.g., Wilson, *Sociobiology: The New Synthesis*; Dawkins, *Selfish Gene*; Barash, *Sociobiology and Behavior*; Freedman, *Human Sociobiology*.

8. S. J. Gould, "Sociobiology and the Theory of Natural Selection," in *Sociobiology: Beyond Nature/Nurture*, ed. G. W. Barlow and J. Silverberg (Boulder, Colo.: Westview Press, 1980), 257–69.

9. V. Reynolds et al., eds., *The Sociobiology of Ethnocentrism* (London: Croom Helm, 1987), xv.

10. V. Reynolds, "Sociobiology and Race Relations," in ibid., 212.

11. Gould, *Mismeasure of Man*.

12. See Hitler, *Mein Kampf*.

13. Proctor, *Racial Hygiene*.

14. Ibid., 123.

15. Hitler, *Mein Kampf*, 441.

16. See Proctor, *Racial Hygiene*, 125.

17. See C. B. Flood, *Hitler: The Path to Power* (Boston: Houghton Mifflin, 1989), 177.

18. Quoted in ibid., 177.

19. Ibid.

20. Time-Life Books, *The New Order* (Alexandria, Va.: Time-Life, 1989).

21. Ibid.

22. Quoted in ibid., 118.

23. Proctor, *Racial Hygiene*, 122.

24. Translated in ibid., 124, from sections by Lenz in Erwin Baur, Eugen Fischer, and Fritz Lenz, *Grundriss der menschlichen Erblichkeitslehre und Rassenhygiene*, vol. 1, 3rd ed. (Munich: J. F. Lehmann, 1927).

25. Ibid.

26. Ibid., 125.

27. Freedman, *Human Sociobiology*.

28. Dawkins, *Selfish Gene*.

29. Wilson, *Sociobiology: The New Synthesis*, 575.

30. Ibid., 3.

31. Wilson, "Human Decency."

32. Wilson, *Sociobiology: The New Synthesis*, 554.

33. Ibid., 549.

34. Ibid., 554.

35. Dawkins, *Selfish Gene*, 21.

36. Wilson, *Sociobiology: The New Synthesis*, 3.

37. Dawkins, *Selfish Gene*, 2

38. Translated by Proctor, *Racial Hygiene*, 51, from sections by Lenz in Baur et al., *Grundriss der menschlichen Erblichkeitslehre und Rassenhygiene*, vol. 1.

39. Dawkins, *Selfish Gene*, 177–78.
40. Translated by Proctor, *Racial Hygiene*, 51.
41. Freedman, *Human Sociobiology.*
42. Ibid., 72.
43. Ibid., 74.
44. Ibid., 75.
45. Ibid., 70.
46. Ibid.
47. Ibid., 194.
48. Quoted in Flood, *Hitler: The Path to Power*, 162.
49. Translated by Proctor, *Racial Hygiene*, 52; from Baur et al., *Erblichkeitslehre.*
50. Ibid.
51. Translated in ibid.
52. See Rushton, "An Evolutionary Theory of Health"; Rushton, "Race Differences."
53. See Gasman, *Scientific Origins of Natural Socialism.*
54. Reynolds, "Sociobiology and Race Relations," 213.
55. Ibid.
56. Rushton, "Race Differences," 1009.
57. Stein, "Biological Bases of Ethnocentrism."
58. E. Haeckel, *The History of Creation: On the Development of the Earth and Its Inhabitants by the Action of Natural Causes* (New York: Appleton, 1876), 365.
59. Dawkins, *Selfish Gene.*
60. Rushton, "Do r/K Reproductive Strategies Apply?" 339.
61. See, e.g., D. C. Johanson and M. A. Edey, *Lucy: The Beginnings of Humankind* (New York: Simon & Schuster, 1981); Wilson, *Sociobiology.*
62. Rushton, "Race Differences," 1018.
63. Dawkins, *Selfish Gene*, 125–26.
64. Wilson, *Sociobiology: The New Synthesis*, 575. On Schallmayer's fear, see Wilhelm Schallmayer, "Gobineaus Rassenwerk und die moderne Gobineauschule," *Zeitschrift für Sozialwissenschaft* 1 (1910): 553–72.
65. Rushton, "Evolutionary Theory"; Freedman, *Human Sociobiology.* Here is the citation for the anonymous surgeon's journal as it appeared in Rushton's 1988 article: "A French Army Surgeon (1898/1972) *Untrodden Fields of Anthropology* (2 Vols.). Charles Carrington, Paris, France Publishers (Reprinted Huntington, New York: Robert E. Krieger Publishing Company)" [*sic*].
66. See Gould, *Mismeasure of Man.*
67. Rushton, "Evolutionary Theory"; Rushton, "Race Differences"; Rushton, "Do r/K Reproductive Strategies Apply?"
68. Rushton, "Race Differences," 1016.
69. Rushton, quoted in *The Globe and Mail*, February 4, 1989.
70. See Konner, *Tangled Wing*; Wilson, *Sociobiology: The New Synthesis*; Wilson, "Academic Vigilantism."
71. Wilson, "Academic Vigilantism," 302.
72. See, e.g., van den Berghe, "Why Most Sociologists . . ."; Wilson, "For Sociobiology"; Wilson, "Academic Vigilantism."
73. Wilson, "A Consideration of the Genetic Foundation."
74. Wilson, "Academic Vigilantism."

CHAPTER 6

1. See Schneirla, "Interrelationships of the Innate"; Schneirla, "Concept of Development"; Tobach, "Some Evolutionary Aspects of Human Gender"; E. Tobach, "The Meaning of

Cryptanthroparion," in *Genetics, Environment, and Behavior*, ed. L. Ehrman et al. (New York: Academic Press, 1972), 37–68; Tobach, "Evolutionary Aspects of the Activity"; Tobach and Schneirla, "The Biopsychology of Social Behavior."

2. See R. Plomin, J. C. DeFries, and G. E. McClearn, *Behavioral Genetics: A Primer* (San Francisco: Freeman, 1980); L. Luzzatto and P. Goodfellow, "A Simple Disease with No Cure," *Nature* 337 (1989): 17–18.

3. See R. M. Lerner, *Human Plasticity*; Luzzatto and Goodfellow, "Simple Disease with No Cure."

4. E. Tobach, Personal communication, November 1990.

5. Lewontin, "On Constraints and Adaptation," 244.

6. H. Katchadourian, *The Biology of Adolescence* (San Francisco: Freeman, 1977).

7. See Gottlieb, "Experiential Canalization"; Hirsch, "Behavior-Genetic Analysis"; R. M. Lerner, *Concepts and Theories*; R. M. Lerner, "Nature, Nurture"; R. M. Lerner, "Dynamic Interactional Concept"; G. E. McClearn, "Evolution and Genetic Variability," in *Developmental Plasticity: Behavioral and Biological Aspects of Variations in Development*, ed. E. S. Golin (New York: Academic Press, 1981), 3–31.

8. W. F. Bodmer and L. L. Cavalli-Sforza, *Genetics, Evolution, and Man* (San Francisco: Freeman, 1976); C. Stern, *Principles of Human Genetics*, 3rd ed. (San Francisco: Freeman, 1973).

9. McClearn, "Evolution and Genetic Variability," 19.

10. Hirsch, "Behavior-Genetic Analysis."

11. Bodmer and Cavalli-Sforzi, *Genetics, Evolution, and Man.*

12. McClearn, "Evolution and Genetic Variability," 19.

13. Ibid., 26.

14. See F. Jacob and J. Monod, "On the Regulation of Gene Activity," *Cold Spring Harbor Symposia on Quantitative Biology* 26 (1961): 193–209; G. E. McClearn, "Genetic Influences on Behavior and Development," in *Carmichael's Manual of Child Psychology*, vol. 1, ed. P. Mussen (New York: Wiley, 1970), 39–76; McClearn, "Evolution and Genetic Variability"; K. W. Schaie et al., *Developmental Human Behavior Genetics* (Lexington, Mass.: Heath, 1975).

15. McClearn, "Evolution and Genetic Variability," 26.

16. See ibid.

17. See R. M. Lerner, "Nature, Nurture"; R. M. Lerner, "Changing Organism-Context Relations as the Basic Process of Development: A Developmental Contextual Perspective," *Developmental Psychology* 27 (1991): 27–32; Lerner and Spanier, *Adolescent Development*; Lewontin et al., "On Constraints and Adaptation."

18. See R. M. Lerner, "Nature, Nurture"; R. M. Lerner, "Dynamic Interactional Concept"; R. M. Lerner, *Concepts and Theories*; R. M. Lerner, "A Life-Span Perspective for Early Adolescence," in *Biological-Psychosocial Interactions in Early Adolescence*, ed. R. M. Lerner and T. T. Foch (Hillsdale, N.J.: Erlbaum, 1987), 1–6; R. M. Lerner, "Changing Organism-Context Relations."

19. Gottlieb, "Experiential Canalization," 5.

20. See Tobach, "Some Evolutionary Aspects of Human Gender"; Tobach, "The Meaning of Cryptanthroparion"; Tobach, "Evolutionary Aspects"; Tobach and Schneirla, "Biopsychology of Social Behavior."

21. Gottlieb, "Experiential Canalization," 5, 8.

22. See Tobach, "The Meaning of Cryptanthroparion"; Tobach, "Evolutionary Aspects"; Tobach and Greenberg, "Schneirla's Contribution."

23. Anastasi, "Heredity, Environment."

24. Tobach, "The Meaning of Cryptanthroparion"; Tobach, "Evolutionary Aspects"; Tobach and Greenberg, "Schneirla's Contribution."

25. Gottlieb, "Experiential Canalization."

26. Ibid., 9.

27. Tobach, Personal communication, November 1990.

28. N. Hunt, *The World of Nigel Hunt: The Diary of a Mongoloid Youth* (New York: Garrett Publications, 1967).

29. See R. M. Lerner, *Human Plasticity.*

30. See L. L. Uphouse and J. Bonner, "Preliminary Evidence for the Effects of Environmental Complexity on Hybridization of Rat Brain RNA to Rat Brain DNA," *Developmental Psychobiology* 8 (1975): 171–78.

31. See L. D. Grouse et al., "Sequence Diversity Studies of Rat Brain RNA: Effects of Environmental Complexity and Rat Brain RNA Diversity," *Journal of Neurochemistry* 30 (1978): 191–203.

32. L. D. Grouse, B. K. Schrier, and P. G. Nelson, "Effect of Visual Experience on Gene Expression During the Development of Stimulus Specificity in Cat Brain," *Experimental Neurology* 64 (1979): 354–64.

33. Anastasi, "Heredity, Environment"; Hirsch, "Behavior-Genetic Analysis"; R. M. Lerner, *Concepts and Theories*; McGuire and Hirsch, "General Intelligence."

34. See, e.g., Hirsch, "Behavior-Genetic Analysis"; McGuire and Hirsch, "General Intelligence."

35. Hirsch, "Behavior-Genetic Analysis."

36. McGuire and Hirsch, "General Intelligence," 25.

37. Ibid., 26.

38. U. Bronfenbrenner, *The Ecology of Human Development* (Cambridge, Mass.: Harvard University Press, 1979); Tobach, "Evolutionary Aspects"; Tobach and Greenberg, "Schneirla's Contribution."

CHAPTER 7

1. See Lorenz, *On Aggression.*

2. Tobach, "Evolutionary Aspects."

3. See Schneirla, "Concept of Development."

4. R. C. Lewontin and R. Levins, "Evolution," in *Encyclopedia*, vol. 5 (Torino, Italy: Einaudi, 1973), 28.

5. C. O. Lovejoy, "The Origin of Man," *Science* 211 (1981): 341–50.

6. Johanson and Edey, *Lucy*, 325.

7. Lewontin and Levins, "Evolution."

8. Johanson and Edey, *Lucy*, 325–26.

9. R. Q. Bell, "A Reinterpretation of the Direction of Effects in Studies of Socialization," *Psychological Review* 75 (1968): 81–95; A. Thomas et al., *Behavioral Individuality in Early Childhood* (New York: New York University Press, 1963).

10. A. Bandura, *Social Learning Theory* (Englewood Cliffs, N.J.: Prentice-Hall, 1977); W. W. Hartup, "Perspectives on Child and Family Interaction: Past, Present, and Future," in *Child Influences on Marital and Family Interaction*, ed. R. M. Lerner and G. B. Spanier (New York: Academic Press, 1978), 23–45; R. M. Lerner and G. B. Spanier, in ibid., 1–22.

11. J. H. Langlois and C. W. Stephan, "Beauty and the Beast: The Role of Physical Attraction in Peer Relationships and Social Behavior," in *Developmental Social Psychology: Theory and Research*, ed. S. S. Brehm et al. (New York: Oxford University Press, 1981), 152–68; G. T. Sorell and C. A. Nowak, "The Role of Physical Attractiveness as a Contributor to

Individual Development," in *Individuals as Producers of Their Development,* ed. R. M. Lerner and N. A. Busch-Rossnagel (New York: Academic Press, 1981), 389–446; Goldberg, "Premature Birth."

12. R. M. Lerner, S. Iwawaki, and T. Chihara, "Development of Personal Space Schemata Among Japanese Children," *Developmental Psychology* 12 (1976): 466–67.

13. Goldberg, "Premature Birth."

14. J. H. Langlois and A. C. Downs, "Peer Relations as a Function of Physical Attractiveness: The Eye of the Beholder or Behavioral Reality?" *Child Development* 50 (1979): 409–18; R. M. Lerner and J. V. Lerner, "Effects of Age, Sex, and Physical Attractiveness on Child-Peer Relations, Academic Performance, and Elementary School Adjustment," *Developmental Psychology* 13 (1977): 585–90.

15. W. Friedrich and J. Boriskin, "The Role of the Child in Abuse: A Review of the Literature," *American Journal of Orthopsychiatry* 7 (1976): 306–13.

16. A. Thomas and S. Chess, *Temperament and Development* (New York: Brunner/Mazel, 1977).

17. S. B. Crockenberg, "Infant Irritability, Mother Responsiveness, and Social Support Influences on the Security of Infant-Mother Attachment," *Child Development* 52 (1981): 857–65; C. Garcia Cole, J. Kagan, and S. J. Reznick, "Behavioral Inhibition in Young Children," *Child Development* 55 (1984): 1005–9.

18. J. E. Bates, "The Concept of Difficult Temperament," *Merrill-Palmer Quarterly* 26 (1980): 299–319; A. Thomas, S. Chess, and S. J. Korn, "The Reality of Difficult Temperament," *Merrill-Palmer Quarterly* 28 (1982): 1–20.

19. S. Chess and A. Thomas, *The Origins and Evolution of Behavior Disorders: Infancy to Early Adult Life* (New York: Brunner/Mazel, 1984).

20. Piaget, *Psychology of Intelligence;* Piaget, "Piaget's Theory."

21. See J. Block, "Assimilation, Accommodation, and the Dynamics of Personality Development," *Child Development* 53 (1982): 281–95; J. H. Block and J. Block, "The Role of Ego-Control and Ego-Resiliency in the Organization of Behavior," in *The Minnesota Symposia on Child Psychology,* vol. 13, ed. W. A. Collins (Hillsdale, N.J.: Erlbaum, 1980), 39–101.

22. R. M. Lerner, *Human Plasticity.*

23. See Thomas and Chess, *Temperament and Development;* J. V. Lerner and R. M. Lerner, "Temperament and Adaptation Across Life: Theoretical and Empirical Issues," in *Life-Span Development and Behavior,* vol. 5, ed. P. B. Baltes and O. G. Brim, Jr. (New York: Academic Press, 1983), 197–230.

24. Thomas and Chess, *Temperament and Development;* A. Thomas and S. Chess, *The Dynamics of Psychological Development* (New York: Brunner/Mazel, 1981); A. Thomas and S. Chess, "The Role of Temperament in the Contributions of Individuals to Their Development," in *Individuals as Producers of Their Development,* ed. R. M. Lerner and N. A. Busch-Rossnagel (New York: Academic Press, 1981), 231–55; J. V. Lerner and R. M. Lerner, "Temperament and Adaptation Across Life," 197–230; R. M. Lerner and J. V. Lerner, "Children in Their Contexts: A Goodness of Fit Model," in *Parenting Across the Life Span: Biosocial Dimensions,* ed. J. B. Lancaster et al. (Chicago: Aldine, 1987), 377–404; R. M. Lerner and J. V. Lerner, "Organismic and Social Contextual Bases of Development: The Sample Case of Adolescence," in *Child Development Today and Tomorrow,* ed. W. Damon (San Francisco: Jossey-Bass, 1989), 69–85.

25. W. Mischel, "On the Future of Personality Measurement," *American Psychologist* 32 (1977): 246–54; M. Snyder, "On the Influence of Individuals on Situations," in *Cognition, Social Interaction, and Personality,* ed. N. Cantor and J. F. Kihlstrom (Hillsdale, N.J.: Erlbaum, 1981), 309–29.

26. See D. Bakan, *The Duality of Human Existence* (Chicago: Rand McNally, 1966).

27. See Dawkins, *Selfish Gene;* Konner, *Tangled Wing;* Wilson, *Sociobiology.*
28. See Hirsch, "Behavior-Genetic Analysis."
29. See Tobach, "Evolutionary Aspects."
30. Lewontin et al., *Not in Our Genes,*10–11.
31. J. M. Tanner, *Growth at Adolescence* (Springfield, Ill.: Thomas, 1962); Tanner, "Growing Up"; J. M. Tanner, "Menarche, Secular Trend in Age of," in *Encyclopedia of Adolescence,* ed. R. M. Lerner et al. (New York: Garland, 1991), 637–41; W. A. Marshall and J. M. Tanner, "Puberty," in *Human Growth,* 2nd ed., vol. 2, ed. F. Falkner and J. M . Tanner (New York: Plenum, 1986), 171–209; Tanner, "Menarche."
32. Tanner, "Menarche."
33. Marshall and Tanner, "Puberty."
34. Katchadourian, *Biology of Adolescence,* 84.
35. Ibid.
36. Ibid., 86.
37. Rushton, "Race Differences."
38. J. Hiernaux, "Ethnic Differences in Growth and Development," *Eugenics Quarterly* 15 (1968): 12–21.
39. See, e.g., P. B. Baltes, "Theoretical Propositions of Life-Span Developmental Psychology: On the Dynamics Between Growth and Decline," *Developmental Psychology* 23 (1987): 611–26; G. H. Elder, Jr., *Children of the Great Depression* (Chicago: University of Chicago Press, 1974); J. R. Nesselroade and P. B. Baltes, "Adolescent Personality Development and Historical Change, 1970–1972," *Monographs of the Society for Research in Child Development* 39 (1974); Schaie, "Primary Mental Abilities in Adulthood."
40. Wilson, "Human Decency Is Animal."
41. Jensen, "How Much Can We Boost IQ," 15.
42. R. Herrnstein, *IQ and the Mediocrity* (Boston: Little, Brown, 1973), 221.
43. See Lewontin et al., *Not in Our Genes,* 70.
44. Jensen, "How Much Can We Boost IQ."
45. See Lewontin et al., *Not in Our Genes.*
46. Translated by Proctor, *Racial Hygiene,* 51, from sections by Lenz in Baur et al., *Erblichkeitslehre.*
47. Ibid.
48. Rushton, quoted in *The Globe and Mail,* February 4, 1989. See Chapter 5 for a discussion.
49. Kachigan, *Sexual Matrix,* 105.
50. See Chapter 2.
51. Kachigan, *Sexual Matrix,* 151.
52. Jensen, "How Much Can We Boost IQ," 95.
53. J. Crow, J. V. Neel, and C. Stern, "Racial Studies: Academy States Position on Call for New Research," *Science* 158 (1967): 893.
54. J. Crow, "Genetic Theories and Influences: Comments on the Value of Diversity," *Harvard Educational Review* 39 (1969): 158.
55. J. F. Crow, "Eugenics: Must It Be a Dirty Word?" *Contemporary Psychology* 33 (1988): 10–12; and Humphreys in R. L. Linn, ed., *Intelligence: Measurement, Theory, and Public Policy—Proceedings of a Symposium in Honor of Lloyd G. Humphreys* (Urbana: University of Illinois Press, 1989).
56. R. C. Birkel, R. M. Lerner, and M. A. Smyer, "Applied Developmental Psychology as an Implementation of a Life Span View of Human Development," *Journal of Applied Developmental Psychology* 10 (1989): 430.
57. Ibid.

58. See ibid., 431.

59. See ibid.

60. Ibid.

61. Ibid., 432.

62. I. Lazar et al., "Lasting Effects of Early Education: A Report from the Consortium for Longitudinal Studies," *Monographs of the Society for Research in Child Development* 47 (1982).

63. See Baltes, "Theoretical Propositions"; P. B. Baltes, H. W. Reese, and L. P. Lipsitt, "Life-Span Developmental Psychology," *Annual Review of Psychology* 31 (1980): 65–110.

64. S. B. Sarason, "Jewishness, Blackishness, and the Nature-Nurture Controversy," *American Psychologist* 28 (1973): 962–71.

65. G. Y. Steiner, *The Futility of Family Policy* (Washington, D.C.: The Brookings Institute, 1981).

66. R. M. Lerner and J. G. Tubman, "Plasticity in Development: Ethical Implications," in *Ethics in Applied Developmental Psychology,* ed. C. B. Fisher and W. W. Tryon (Norwood, N.J.: Ablex, 1990).

67. Birkel et al., "Applied Developmental Psychology," 433.

68. Baltes, "Theoretical Propositions."

EPILOGUE

1. Lifton, *Nazi Doctors,* 13.

2. Lewontin, "On Constraints and Adaptation."

3. Lifton, *Nazi Doctors,* 417.

SELECTED BIBLIOGRAPHY

Allen, E.; Beckwith, B.; Beckwith, J.; Chorover, S.; Culver, D.; Duncan, M.; Gould, S. J.; Hubbard, R.; Inouye, H.; Leeds, A.; Lewontin, R.; Madansky, C.; Miller, L.; Pyeritz, R.; Rosenthal, M.; and Schreier, H. "Against 'Sociobiology.' " *New York Review of Books*, November 13, 1975, pp. 182, 184–86.

Alper, J.; Beckwith, J.; and Miller, L. G. "Sociobiology Is a Political Issue." In A. L. Caplan, ed., *The Sociobiology Debate*, 476–88. New York: Harper & Row, 1978.

Anastasi, A. "Heredity, Environment, and the Question 'How?' " *Psychological Review* 65 (1958): 197–208.

Atz, J. W. "The Application of the Idea of Homology to Behavior." In L. R. Aronson, E. Tobach, D. S. Lehrman, and J. S. Rosenblatt, eds., *Development and Evolution of Behavior: Essays in Memory of T. C. Schneirla*, 53–74. San Francisco: Freeman, 1970.

Bakan, D. *The Duality of Human Existence*. Chicago: Rand McNally, 1966.

Baltes, P. B. "Theoretical Propositions of Life-Span Developmental Psychology: On the Dynamics Between Growth and Decline." *Developmental Psychology* 23 (1987): 611–26.

Baltes, P. B.; Reese, H. W.; and Lipsitt, L. P. "Life-Span Developmental Psychology." *Annual Review of Psychology* 31 (1980): 65–110.

Bandura, A. *Social Learning Theory*. Englewood Cliffs, N.J.: Prentice-Hall, 1977.

Barash, D. P. *Sociobiology and Behavior*. New York: Elsevier, 1977.

Barlow, G. W. "The Development of Sociobiology: A Biologist's Perspective." In G. W. Barlow and J. Silverberg, eds., *Sociobiology: Beyond Nature/Nurture?*, 3–34. Boulder, Colo.: Westview Press, 1980.

Bateman, A. J. "Intrasexual Selection in Drosophila." *Heredity* 2 (1980): 349–68.

Bates, J. E. "The Concept of Difficult Temperament." *Merrill-Palmer Quarterly* 26 (1980): 299–319.

Bauer, Y. "Genocide: Was It the Nazis' Original Plan?" *Annals of the American Academy of Political and Social Science* 450 (1980): 35–45.

Baur, E.; Fischer, E.; and Lenz, F. *Grundriss der menschlichen Erblichkeitslehre und Rassenhygiene*, vol. 1, 3rd ed. Munich: J. F. Lehmann, 1927.

Bell, R. Q. "A Reinterpretation of the Direction of Effects in Studies of Socialization." *Psychological Review* 75 (1968): 81–95.

Berghe, P. L. van den. "Why Most Sociologists Don't (and Won't) Think Evolutionarily." *Sociological Forum* 5 (1990): 173–85.

Berghe, P. L. van den, and Barash, D. P. "Inclusive Fitness and Human Family Structure." *American Anthropologist* 79 (1977): 809–23.

Bertalanffy, L. von. "Chance or Law." In A. Koestler, ed., *Beyond Reductionism*. London: Hitchinson, 1969.

Billig, M. *Fascists: A Social Psychological View of the National Front*. New York: Academic Press, 1978.

Binding, K., and Hoche, A. *Die Freigabe der Vernichtung lebensunwerten Lebens* (The Sanctioning of the Destruction of Lives Unworthy to Be Lived). Leipzig: F. Meiner, 1920.

Birkel, R. C.; Lerner, R. M.; and Smyer, M. A. "Applied Developmental Psychology as an Implementation of a Life Span View of Human Development." *Journal of Applied Developmental Psychology* 10 (1989): 425–45.

Bitterman, M. E. "The Comparative Analysis of Learning." *Science* 18 (1975): 699–709.

———. "Phyletic Differences in Learning." *American Psychologist* 20 (1965): 396–410.

Block, J. "Assimilation, Accommodation, and the Dynamics of Personality Development." *Child Development* 53 (1982): 281–95.

Block, J. H., and Block, J. "The Role of Ego-Control and Ego-Resiliency in the Organization of Behavior." In W. A. Collins, ed., *The Minnesota Symposia on Child Psychology*, vol. 13, pp. 39–101. Hillsdale, N.J.: Erlbaum, 1980.

Bock, W. J. "The Definition and Recognition of Biological Adaptation." *American Zoologist* 20 (1980): 217–27.

———. "A Synthetic Explanation of Macroevolutionary Change—A Reductionistic Approach." *Bulletin of the Carnegie Museum of Natural History* 13 (1979): 20–69.

———. "The Use of Adaptive Characters in Avian Classification." *Proceedings of the XIV International Ornithology Congress*, July 1966, Oxford, England.

Bodmer, W. F., and Cavalli-Sforza, L. L. *Genetics, Evolution, and Man*. San Francisco: Freeman, 1976.

Brennecke, F., ed. *Vom deutschen Volk und seinem Lebensraum: Handbuch für die Schulungsarbeit in der Hitler Jugend*. Munich: Zentralverlag der NSDAP, 1937.

Brim, O. G., Jr., and Kagan, J. "Constancy and Change: A View of the Issues." In O. G. Brim, Jr., and J. Kagan, eds., *Constancy and Change in Human Development*, 1–25. Cambridge, Mass.: Harvard University Press, 1980.

Brim, O. G., Jr., and Kagan, J., eds. *Constancy and Change in Human Development*. Cambridge, Mass.: Harvard University Press, 1980.

Bronfenbrenner, U. *The Ecology of Human Development*. Cambridge, Mass.: Harvard University Press, 1980.

Brooks-Gunn, J., and Petersen, A. C., eds. *Girls at Puberty: Biological and Psychosocial Perspectives*. New York: Plenum, 1983.

Brown, D. *Bury My Heart at Wounded Knee.* New York: Washington Square Press, 1981.
———. *The Westerners.* New York: Holt, Rinehart & Winston, 1974.
Browning, C. R. "The Decision Concerning the Final Solution." In F. Furet, ed., *Unanswered Questions: Nazi Germany and the Genocide of the Jews,* 96–118. New York: Schocken Books, 1989.
Brownmiller, S. *Against Our Will: Men, Women, and Rape.* New York: Simon & Schuster, 1975.
Bullock, A. *Hitler: A Study in Tyranny.* Revised edition. New York: Harper & Row, 1962.
Bullock, D. "Methodological Heterogeneity and the Anachronistic Status of ANOVA in Psychology." *Behavioral and Brain Sciences* 13 (1990): 122–23.
Caplan, A. L. "A Critical Examination of Current Sociobiological Theory: Adequacy and Implications." In G. W. Barlow and J. Silverberg, eds., *Sociobiology: Beyond Nature/Nurture?,* 97–121. Boulder, Colo.: Westview Press, 1980.
Caplan, A. L., ed. *The Sociobiology Debate.* New York: Harper & Row, 1978.
Chagnon, N. A. *Yanomamo: The Fierce People.* New York: Holt, Rinehart & Winston, 1972.
Chess, S., and Thomas, A. *The Origins and Evolution of Behavior Disorders: Infancy to Early Adult Life.* New York: Brunner/Mazel, 1984.
Chorover, S. L. *From Genesis to Genocide.* Cambridge, Mass.: MIT Press, 1979.
Cohn, N. *Warrant for Genocide.* Atlanta, Ga.: Scholars Press, 1981.
Cox, V. "A Prize for the Goose Father." *Human Behavior* 3 (1974): 17–22.
Crockenberg, S. B. "Infant Irritability, Mother Responsiveness, and Social Support Influences on the Security of Infant-Mother Attachment." *Child Development* 52 (1981): 857–65.
Crow, J. F. "Eugenics: Must It Be a Dirty Word?" *Contemporary Psychology* 33 (1988): 10–12.
———. "Genetic Theories and Influences: Comments on the Value of Diversity." *Harvard Educational Review* 39 (1969): 301–9.
Crow, J. F.; Neel, J. V.; and Stern, C. "Racial Studies: Academy States Position on Call for New Research." *Science* 158 (1967): 892–93.
Darwin, C. *The Descent of Man,* 2nd ed. London: Murray, 1922.
———. *On the Origin of Species by Means of Natural Selection; or, The Preservation of Favored Races in the Struggle for Life.* London: Murray, 1859.
Dawidowicz, L. S. *The War Against the Jews, 1933–1945.* New York: Holt, Rinehart & Winston, 1975.
Dawkins, R. *The Selfish Gene.* New York: Oxford University Press, 1976.
Dennen, J.M.G. ven der. "Ethnocentrism and In-Group/Out-Group Differentiation: A Review and Interpretation of the Literature." In V. Reynolds, V. Falger, and I. Vine, eds., *The Sociobiology of Ethnocentrism,* 1–47. London: Croom Helm, 1987.
Dunbar, R.I.M. "Sociobiological Explanations and the Evolution of Ethnocentrism." In V. Reynolds, V. Falger, and I. Vine, eds., *The Sociobiology of Ethnocentrism,* 48–59. London: Croom Helm, 1987.
Eisenberg, L. "On the Human Nature of Human Nature." *Science* 176 (1972): 123–28.
Elder, G. H., Jr. *Children of the Great Depression.* Chicago: University of Chicago Press, 1974.
Erikson, E. H. "Identity and the Life-Cycle." *Psychological Issues* 1 (1959): 18–164.

————. *Identity, Youth, and Crisis.* New York: Norton, 1968.

Evans, R. I. "Interview with Konrad Lorenz." *Psychology Today* 8 (1974): 26ff.

Evans, R. I., ed. *Konrad Lorenz: The Man and His Ideas.* New York: Harcourt Brace Jovanovich, 1975.

Falconer, D. S. *Introduction to Quantitative Genetics.* New York: Ronald Press, 1960.

Feldman, M. W., and Lewontin, R. C. "The Heritability Hang-Up." *Science* 190 (1975): 1163–68.

Fischer, E. *Der völkische Staat, biologisches Geschehen.* Berlin: Junker & Dünnhaupt, 1933.

Flohr, H. "Biological Bases of Social Prejudices." In V. Reynolds, V. Falger, and I. Vine, eds., *The Sociobiology of Ethnocentrism,* 190–207. London: Croom Helm, 1987.

Flood, C. B. *Hitler: The Path to Power.* Boston: Houghton Mifflin, 1989.

Freedman, D. G. *Human Sociobiology: A Holistic Approach.* New York: The Free Press, 1979.

Freud, S. *The Ego and the Id.* London: Hogarth, 1923.

Friedlander, S. "From Anti-Semitism to Extermination." In F. Furet, ed., *Unanswered Questions: Nazi Germany and the Genocide of the Jews,* 3–31. New York: Schocken Books, 1989.

Friedrich, W., and Boriskin, J. "The Role of the Child in Abuse: A Review of the Literature." *American Journal of Orthopsychiatry* 7 (1976): 306–13.

Furet, F., ed. *Unanswered Questions: Nazi Gemany and the Genocide of the Jews.* New York: Schocken Books, 1989.

Garcia Coll, C.; Kagan, J.; and Resnick, S. J. "Behavioral Inhibition in Young Children." *Child Development* 55 (1984): 1005–9.

Gasman, D. *The Scientific Origins of Natural Socialism: Social Darwinism in Ernst Haeckel and the German Monist League.* New York: Elsevier, 1971.

Gesell, A. L. "The Ontogenesis of Infant Behavior." In L. Carmichael, ed., *Manual of Child Psychology,* 295–331. New York: Wiley, 1946.

Gilbert, M. *The Holocaust: A History of the Jews of Europe During the Second World War.* New York: Holt, Rinehart & Winston, 1985.

Goldberg, S. "Premature Birth: Consequences for the Parent-Infant Relationship." *American Scientist* 67 (1979): 214–70.

Gollin, E. S. "Development and Plasticity." In E. S. Gollin, ed., *Developmental Plasticity: Behavioral and Biological Aspects of Variations in Development,* 231–331. New York: Academic Press, 1981.

Gottlieb, G. "The Experiential Canalization of Behavioral Development." *Developmental Psychology* 27 (1991): 4–13, 35–39.

Gould, S. J. *The Mismeasure of Man.* New York: Norton, 1981.

————. "Sociobiology and the Theory of Natural Selection." In G. W. Barlow and J. Silverberg, eds., *Sociobiology: Beyond Nature/Nurture,* 257–69. Boulder, Colo.: Westview Press, 1980.

Gould, S. J., and Lewontin, R. C. "The Spandrels of San Marco and the Panglossian Paradigm: A Critique of the Adaptationist Programme." In J. Maynard Smith and R. Holliday, eds., *The Evolution of Adaptation by Natural Selection,* 581–98. London: Royal Society of London, 1979.

Gould, S., and Vrba, E. "Exaptation: A Missing Term in the Science of Form." *Paleobiology* 8 (1982): 4–15.

Grouse, L. D.; Schrier, B. K.; Bennett, E. L.; Rosenzweig, M. R.; and Nelson, P. G.

"Sequence Diversity Studies of Rat Brain RNA: Effects of Environmental Complexity and Rat Brain RNA Diversity." *Journal of Neurochemistry* 30 (1978): 191–203.

Grouse, L. D.; Schrier, B. K.; and Nelson, P. G. "Effect of Visual Experience on Gene Expression During the Development of Stimulus Specificity in Cat Brain." *Experimental Neurology* 64 (1979): 354–64.

Grzimek, B., ed. *Animal Life Encyclopedia*, vol. 7. New York: Van Nostrand Reinhold, 1972.

Haeckel, E. *The History of Creation; or, The Development of the Earth and Its Inhabitants by the Action of Natural Causes*. New York: Appleton, 1876.

———. *The Wonders of Life*. New York: Harper, 1905.

Hartup, W. W. "Perspectives on Child and Family Interaction: Past, Present, and Future." In R. M. Lerner and G. B. Spanier, eds., *Child Influences on Marital and Family Interaction: A Life-Span Perspective*, 23–45. New York: Academic Press, 1978.

Hebb, D. O. "A Return to Jensen and His Social Critics." *American Psychologist* 25 (1970): 568.

Herrnstein, R. *IQ and the Meritocracy*. Boston: Little, Brown, 1973.

Hiernaux, J. "Ethnic Differences in Growth and Development." *Eugenics Quarterly* 15 (1968): 12–21.

Hilberg, R. *The Destruction of the European Jews*. Chicago: Quadrangle Books, 1961.

Hirsch, J. "Behavior-Genetic Analysis and Its Biosocial Consequences." *Seminars in Psychiatry* 2 (1970): 89–105.

———. "Correlation, Causation, and Careerism." *European Bulletin of Cognitive Psychology* 10 (1990): 647–52.

———. "A Nemesis for Heritability Estimation." *Behavioral and Brain Sciences* 13 (1990): 137–38.

———. "Obfuscation of Interaction Through Incantation." Manuscript. University of Illinois at Urbana, 1990.

———. Review of *Sociobiology*, by E. O. Wilson. *Animal Behavior* 24 (1976): 707–9.

———. "To 'Unfrock the Charlatans.'" *SAGE Race Relations Abstracts* 6 (1981): 1–65.

Hirsch, J.; McGuire, T. R.; and Vetta, A. "Concepts of Behavior Genetics and Misapplications to Humans." In J. S. Lockard, ed., *The Evolution of Human Social Behavior*, 215–38. New York: Elsevier, 1980.

Hitler, A. *Mein Kampf*. Trans. R. Manheim. Boston: Houghton Mifflin, 1925 (1927/ 1943).

Hovannisian, R. G., ed. *The Armenian Genocide in Perspective*. New Brunswick, N.J.: Transaction Books, 1986.

Hunt, N. *The World of Nigel Hunt: The Diary of a Mongoloid Youth*. New York: Garrett, 1967.

Irwin, C. J. "A Study in the Evolution of Ethnocentrism." In V. Reynolds, V. Falger, and I. Vine, eds., *The Sociobiology of Ethnocentrism*, 131–56. London: Croom Helm, 1987.

Jacob, F., and Monod, J. "On the Regulation of Gene Activity." *Cold Spring Harbor Symposia on Quantitative Biology* 26 (1961): 193–209.

Jensen, A. R. "How Much Can We Boost IQ and Scholastic Achievement?" *Harvard Educational Review* 39 (1969): 1–123.

Johanson, D. C., and Edey, M. A. *Lucy: The Beginnings of Humankind.* New York: Simon & Schuster, 1981.

Kachigan, S. K. *The Sexual Matrix.* New York: Radius Press, 1990.

Kalikow, T. J. "Konrad Lorenz's 'Brown Past': A Reply to Alec Nisbett." *Journal of the History of the Behavioral Sciences* 14 (1978): 173–79.

———. "Konrad Lorenz's Ethological Theory: Explanation and Ideology, 1938–1943." *Journal of the History of Biology* 16 (1983): 39–73.

———. Review of *Civilized Man's Eight Deadly Sins*, by Konrad Lorenz. *Philosophy of the Social Sciences* 8 (1978): 99–108.

Katchadourian, H. *The Biology of Adolescence.* San Francisco: Freeman, 1977.

Kempthorne, O. "How Does One Apply Statistical Analysis to Our Understanding of the Development of Human Relationships?" *Behavioral and Brain Sciences* 13 (1990): 138–39.

Kevles, D. J. *In the Name of Eugenics.* New York: Knopf, 1985.

Kinsey, A. C., et al. *Sexual Behavior in the Human Female.* Philadelphia: Saunders, 1953.

Kitcher, P. *Vaulting Ambition.* Cambridge, Mass.: MIT Press, 1985.

Koford, C. B. "Group Relations in an Island Colony of Rhesus Monkeys." In C. H. Southwick, ed., *Primate Social Behavior*, 136–52. Princeton: Van Nostrand, 1963.

Konner, M. *The Tangled Wing.* New York: Holt, Rinehart & Winston, 1982.

Kuhn, T. S. *The Structure of Scientific Revolutions*, 2nd ed. Chicago: University of Chicago Press, 1970.

Langlois, J. H., and Downs, A. C. "Peer Relations as a Function of Physical Attractiveness: The Eye of the Beholder or Behavioral Reality?" *Child Development* 50 (1979): 409–18.

Langlois, J. H., and Stephan, C. W. "Beauty and the Beast: The Role of Physical Attraction in Peer Relationships and Social Behavior." In S. S. Brehm, S. M. Kassin, and S. X. Gibbons, eds., *Developmental Social Psychology: Theory and Research*, 152–68. New York: Oxford University Press, 1981.

Lazar, I.; Darlington, R.; Murray, H.; Royce, J.; and Snipper, A. "Lasting Effects of Early Education: A Report from the Consortium for Longitudinal Studies." *Monographs of the Society for Research in Child Development* 47, serial no. 195, nos. 2–3 (1982).

Lehrman, D. S. "Semantic and Conceptual Issues in the Nature-Nurture Problem." In L. R. Aronson, E. Tobach, D. S. Lehrman, and J. S. Rosenblatt, eds., *Development and Evolution of Behavior: Essays in Memory of T. C. Schneirla*, 17–52. San Francisco: Freeman, 1970.

Lenz, F. "Alfred Ploetz zum 70. Geburtstag am 22 August." *Archiv für Rassen- und Gesellschaftsbiologie* 24 (1930): xiv.

———. Review of Eugen Fischer article in the German "Dictionary of the Sciences." *Archiv für Rassen- und Gesellschaftsbiologie* 10 (1913): 369.

———. *Über die krankhaften Erbanlagen des Mannes und die Bestimmung des Geschlechts beim Menschen.* Medical dissertation, University of Jena, 1912.

Lerner, J. V. "The Import of Temperament for Psychosocial Functioning: Tests of a 'Goodness of Fit' Model." *Merrill-Palmer Quarterly* 30 (1984): 177–88.

Lerner, J. V., and Lerner, R. M. "Temperament and Adaptation Across Life: Theoretical and Empirical Issues." In P. B. Baltes and O. G. Brim, Jr., eds., *Life-Span Development and Behavior*, vol. 5, pp. 197–230. New York: Academic Press, 1983.

Lerner, R. M. "Changing Organism-Context Relations as the Basic Process of Development: A Developmental Contextual Perspective." *Developmental Psychology* 27 (1991): 27–32.
———. *Concepts and Theories of Human Development*. Reading, Mass.: Addison-Wesley, 1976. Second edition, Random House, 1986.
———. "A Dynamic Interactional Concept of Individual and Social Relationship Development." In R. Burgess and T. Huston, eds., *Social Exchange in Developing Relationships*, 271–305. New York: Academic Press, 1979.
———. "A Life-Span Perspective for Early Adolescence." In R. M. Lerner and T. T. Foch, eds., *Biological-Psychosocial Interactions in Early Adolescence: A Life-Span Perspective*, 1–6. Hillsdale, N.J.: Erlbaum, 1987.
———. "Nature, Nurture, and Dynamic Interactionism." *Human Development* 21 (1978): 1–20.
———. *On the Nature of Human Plasticity*. New York: Cambridge University Press, 1984.
Lerner, R. M., and Foch, T. T., eds. *Biological-Psychosocial Interactions in Early Adolescence*. Hillsdale, N.J.: Erlbaum, 1987.
Lerner, R. M.; Hultsch, D. F.; and Dixon, R.A. "Contextualism and the Character of Developmental Psychology in the 1970s." *Annals of the New York Academy of Sciences* 412 (1983): 101–28.
Lerner, R. M.; Iwawaki, S.; and Chihara, T. "Development of Personal Space Schemata Among Japanese Children." *Developmental Psychology* 12 (1976): 466–67.
Lerner, R. M., and Kauffman, M. B. "The Concept of Development in Contextualism." *Developmental Review* 5 (1985): 309–33.
Lerner, R. M., and Lerner, J. V. "Children in Their Contexts: A Goodness of Fit Model." In J. B. Lancaster, J. Altmann, A. S. Rossi, and L. R. Sherrod, eds., *Parenting Across the Life Span: Biosocial Dimensions*, 377–404. Chicago: Aldine, 1987.
———. "Effects of Age, Sex, and Physical Attractiveness on Child-Peer Relations, Academic Performance, and Elementary School Adjustment." *Developmental Psychology* 13 (1977): 585–90.
———. "Organismic and Social Contextual Bases of Development: The Sample Case of Adolescence." In W. Damon, ed., *Child Development Today and Tomorrow*, 69–85. San Francisco: Jossey-Bass, 1989.
Lerner, R. M., and Spanier, G. B. *Adolescent Development: A Life-Span Perspective*. New York: McGraw-Hill, 1980.
———. "A Dynamic Interactional View of Child and Family Development." In R. M. Lerner and G. B. Spanier, eds., *Child Influences on Marital and Family Interaction: A Life-Span Perspective*, 1–22. New York: Academic Press, 1978.
Lerner, R. M., and Tubman, J. "Plasticity in Development: Ethical Implications." In C. B. Fisher and W. W. Tryon, eds., *Ethics in Applied Developmental Psychology*, 113–31. Norwood, N.J.: Ablex, 1990.
Lewontin, R. C. "On Constraints and Adaptation. *Behavioral and Brain Sciences* 4 (1981): 244–45.
Lewontin, R. C., and Levins, R. "Evolution." In *Encyclopedia*, vol. 5. Torino, Italy: Einaudi, 1973.
Lewontin, R. C.; Rose, S.; and Kamin, L. J. *Not in Our Genes: Biology, Ideology, and Human Nature*. New York: Pantheon Books, 1984.
Lifton, R. J. *The Nazi Doctors: Medical Killing and the Psychology of Genocide*. New York: Basic Books, 1986.

Linn, R. L., ed. *Intelligence: Measurement, Theory, and Public Policy.* Proceedings of a symposium in honor of Lloyd G. Humphreys. Urbana: University of Illinois Press, 1989.

Lorenz, K. "Die angeborenen Formen möglicher Erfahrung." *Zeitschrift für Tierpsychologie* 5 (1943): 235–409.

——. *Civilized Man's Eight Deadly Sins.* New York: Harcourt Brace Jovanovich, 1974.

——. "A Consideration of Methods of Identification of Species-Specific Instinctive Patterns in Birds." In R. Martin, ed. and trans., *Studies in Animal and Human Behavior*, vol. 1, pp. 57–100. 1932. Reprint. Cambridge, Mass.: Harvard University Press, 1970.

——. "Durch Domestikation verursachte Störungen arteigenen Verhaltens." *Zeitschrift für angewandte Psychologie und Charakterkunde* 59 (1940): 2–81.

——. *Evolution and Modification of Behavior.* Chicago: University of Chicago Press, 1965.

——. "Ganzheit und Teil in der tierischen und menschlichen Gemeinschaft." *Studium Generale* 9 (1950): 455–98.

——. "Konrad Lorenz Responds to Donald Campbell." In R. I. Evans, ed., *Konrad Lorenz: The Man and His Ideas*, 119–28. New York: Harcourt Brace Jovanovich, 1975.

——. "Der Kumpan in der Umwelt des Vogels." *Journal für Ornithologie* 83 (1935): 137–215, 289–413.

——. "Letter: Lorenz Clarifies Ideas." *Human Behavior*, September 1974, p. 6.

——. *On Aggression.* New York: Harcourt, Brace & World, 1966.

——. "Psychologie und Stammesgeschichte." In G. Heberer, ed., *Die Evolution der Organismen*, 105–27. Jena: G. Fischer, 1943. Second edition, 1954.

——. "Systematik und Entwicklungsgedanke im Unterricht." *Der Biologe* 9 (1940): 24–36.

——. "Über Ausfallserscheinungen im Instinktverhalten von Haustieren und ihre socialpsychologische Bedeutung." In O. Klemm, ed., *Charakter und Erziehung: 16. Kongress der Deutschen Gesellschaft für Psychologie in Bayreuth*, 139–47. Leipzig: J. A. Barth, 1939.

——. "Über den Begriff der Instinkthandlung." *Folia Biotheoretica* 2 (1937): 17–50.

——. "Über die Bildung des Instinktbegriffes." *Die Naturwissenschaften* 25 (1937): 189–300, 307–8, 324–33.

Lovejoy, C. O. "The Origin of Man." *Science* 211 (1981): 341–50.

Lumsden, C. J., and Wilson, E. O. *Genes, Mind, and Culture.* Cambridge, Mass.: Harvard University Press, 1981.

Luzzatto, L., and Goodfellow, P. "A Simple Disease with No Cure." *Nature* 337 (1989): 17–18.

Marshall, W. A., and Tanner, J. M. "Puberty." In F. Falkner and J. M. Tanner, eds., *Human Growth*, 2nd ed., vol. 2, pp. 171–209. New York: Plenum, 1986.

McClearn, G. E. "Evolution and Genetic Variability." In E. S. Golin, ed., *Developmental Plasticity: Behavioral and Biological Aspects of Variations in Development*, 3–31. New York: Academic Press, 1981.

——. "Genetic Influences on Behavior and Development." In *Carmichael's Manual of Child Psychology*, ed. P. H. Mussen, vol. 1, 39–76. New York: Wiley, 1970.

McGuire, T. R., and Hirsch, J. "General Intelligence and Heritability." In I. C.

Uzgiris and F. Weizmann, eds., *The Structuring of Experience*, 25–72. New York: Plenum, 1977.

Mischel, W. "On the Future of Personality Measurement." *American Psychologist* 32 (1977): 246–54.

Müller-Hill, B. *Murderous Science: Elimination by Scientific Selection of Jews, Gypsies, and Others, Germany 1933–1945*, trans. G. R. Fraser. New York: Oxford University Press, 1988.

Muuss, R. E. "The Philosophical and Historical Roots of Theories of Adolescence." In R. E. Muuss, ed., *Adolescent Behavior and Society: A Book of Readings*, 2nd ed., 4–24. New York: Random House, 1975.

Nesselroade, J. R., and Baltes, P. B. "Adolescent Personality Development and Historical Change, 1970–1972." *Monographs of the Society for Research in Child Development* 39, no. 154 (1974).

Nisbett, A. *Konrad Lorenz: A Biography*. New York: Harcourt Brace Jovanovich, 1977.

Nolte, E. *Three Faces of Fascism: Action Française, Italian Fascism, National Socialism*. New York: Holt, Rinehart & Winston, 1965.

Overton, W. F. "On the Assumptive Base of the Nature-Nurture Controversy: Additive Versus Interactive Conceptions." *Human Development* 16 (1973): 74–89.

———. "The Structure of Developmental Theory." In P. van Geert and L. P. Mos, eds., *Annals of Theoretical Psychology*, vol. 6. New York: Plenum, in press.

Overton, W. F., and Reese, H. W. "Models of Development: Methodological Implications." In J. R. Nesselroade and H. W. Reese, eds., *Life-Span Developmental Psychology: Methodological Issues*, 65–86. New York: Academic Press, 1973.

Pepper, S. C. *World Hypotheses*. Berkeley and Los Angeles: University of California Press, 1942.

Philipp, E. E. "Discussion." In *Law and Ethics of AID and Embryo Transfer*, 66. Ciba Foundation Symposium 17, new series. Amsterdam: Elsevier, Excerpta Medica, North Holland, 1973.

Piaget, J. "Piaget's Theory." In P. H. Mussen, ed., *Carmichael's Manual of Child Psychology*, vol. 1, pp. 703–32. New York: Wiley, 1970.

———. *The Psychology of Intelligence*. New York: Harcourt, Brace, 1950.

Ploetz, A. *The Excellence of Our Race and the Protection of the Weak*. Berlin, 1895.

Plomin, R.; DeFries, J. C.; and McClearn, G. E. *Behavioral Genetics: A Primer*. San Francisco: Freeman, 1980.

Proctor, R. N. *Racial Hygiene: Medicine Under the Nazis*. Cambridge, Mass.: Harvard University Press, 1988.

Reese, H. W., and Overton, W. F. "Models of Development and Theories of Development." In L. R. Goulet and P. B. Baltes, eds., *Life-Span Developmental Psychology: Research and Theory*, 115–45. New York: Academic Press, 1970.

Reynolds, V. "Sociobiology and Race Relations." In V. Reynolds, V. Falger, and I. Vine, eds., *The Sociobiology of Ethnocentrism*, 208–15. London: Croom Helm, 1987.

Reynolds, V.; Falger, V.; and Vine, I. "Introduction." In V. Reynolds, V. Falger, and I. Vine, eds., *The Sociobiology of Ethnocentrism*, xv–xx. London: Croom Helm, 1987.

Reynolds, V.; Falger, V.; and Vine I., eds. *The Sociobiology of Ethnocentrism*. London: Croom Helm, 1987.

Richards, R. J. *Darwin and the Emergence of Evolutionary Theories of Mind and Behavior*. Chicago: University of Chicago Press, 1987.

Rosnow, R., and Georgoudi, M., eds. *Contextualism and Understanding in Behavioral Research*. New York: Praeger, 1986.

Roth, P. *Portnoy's Complaint*. New York: Random House, 1969.

Rushton, J. P. "Do *r/K* Reproductive Strategies Apply to Human Differences?" *Social Biology* 35 (1988): 337–40.

———. "An Evolutionary Theory of Health, Longevity, and Personality: Sociobiology and r/K Reproductive Strategies." *Psychological Reports* 60 (1987): 539–49.

———. "Race Differences in Behavior: A Review and Evolutionary Analysis." *Personality and Individual Differences* 9 (1988): 1009–24.

Sameroff, A. (1975). "Transactional Models in Early Social Relations." *Human Development* 18 (1975): 65–79.

Sarason, S. B. "Jewishness, Blackishness, and the Nature-Nurture Controversy." *American Psychologist* 28 (1973): 962–71.

Sarbin, T. R. "Contextualism: A World View for Modern Psychology." In J. K. Cole and A. W. Lundfield, eds., *Nebraska Symposium on Motivation*, 1976, 1–41. Lincoln: University of Nebraska Press, 1977.

Scarr-Salapatek, S. "Unknowns in the IQ Equation." *Science* 174 (1971): 1223–28.

Schaie, K. W. "The Primary Mental Abilities in Adulthood: An Exploration in the Development of Psychometric Intelligence." In P. B. Baltes and O. G. Brim, Jr., eds., *Life-Span Development and Behavior*, vol. 2, pp. 67–115. New York: Academic Press, 1979.

Schaie, K. W.; Anderson, V. E.; McClearn, G. E.; and Money, J., eds. *Developmental Human Behavior Genetics*. Lexington, Mass.: Heath, 1975.

Schleunes, K. A. "Retracing the Twisted Road." In F. Furet, ed., *Unanswered Questions: Nazi Germany and the Genocide of the Jews*, 54–70. New York: Schocken Books, 1989.

Schneirla, T. C. "The Concept of Development in Comparative Psychology." In D. B. Harris, ed., *The Concept of Development*, 78–108. Minneapolis: University of Minnesota Press, 1957.

———. "Instinct and Aggression." *Natural History* 75 (1966): 16ff.

———. "Interrelationships of the Innate and the Acquired in Instinctive Behavior." In P. P. Grasse, ed., *L'instinct dans le comportement des animaux et de l'homme*, 387–452. Paris: Mason et Cie, 1956.

Shirer, W. L. *The Rise and Fall of the Third Reich*. New York: Simon & Schuster, 1960.

Sicroff, A. A. *Les controverses des statuts de "pureté de sang" en Espagna du 15ᵉ au 17ᵉ siècle*. Paris: Didier, 1960.

Skinner, B. F. *Beyond Freedom and Dignity*. New York: Knopf, 1971.

Snyder, M. "On the Influence of Individuals on Situations." In N. Cantor and J. F. Kihlstrom, eds., *Cognition, Social Interaction, and Personality*, 309–29. Hillsdale, N.J.: Erlbaum, 1981.

Sociobiology Study Group of Science for the People. "Sociobiology: Another Biological Determinism." *BioScience* 26, March 1976.

Sorell, G. T., and Nowak, C. A. "The Role of Physical Attractiveness as a Contributor to Individual Development." In R. M. Lerner and N. A. Busch-Rossnagel, eds., *Individuals as Producers of Their Development: A Life-Span Perspective*, 389–446. New York: Academic Press, 1981.

Spencer, H. *The Principles of Sociology*, vols. 1, 2, and 3. New York: D. Appleton & Co., 1910.

Staub, E. *The Roots of Evil: The Origins of Genocide and Other Group Violence.* New York: Cambridge University Press, 1989.

Stein, G. J. "The Biological Bases of Ethnocentrism, Racism, and Nationalism in National Socialism." In V. Reynolds, V. Falger, and I. Vine, eds., *The Sociobiology of Ethnocentrism*, 251–67. London: Croom Helm, 1987.

Steiner, G. Y. *The Futility of Family Policy.* Washington, D.C.: The Brookings Institute, 1981.

Stern, C. *Principles of Human Genetics*, 3rd ed. San Francisco: Freeman, 1973.

Sunday, S. R., and Tobach, E., eds. *Violence Against Women: A Critique of the Sociobiology of Rape.* New York: Gordian Press, 1985.

Super, C. M., and Harkness, S. "Anthropological Perspectives on Child Development." *New Directions for Child Development*, no. 8. San Francisco: Jossey-Bass, 1980.

———. "Figure, Ground, and Gestalt: The Cultural Context of the Active Individual." In R. M. Lerner and N. A. Busch-Rossnagel, eds., *Individuals as Producers of Their Own Development: A Life-Span Perspective*, 69–86. New York: Academic Press, 1981.

———. "The Infant's Niche in Rural Kenya and Metropolitan America." In L. L. Adler, ed., *Cross-Cultural Research at Issue*, 47–56. New York: Academic Press, 1982.

Tanner, J. M. "Growing Up." *Scientific American* 229 (1973): 34–43.

———. *Growth at Adolescence.* Springfield, Ill.: Thomas, 1962.

———. "Menarche, Secular Trend in Age of." In R. M. Lerner, A. C. Petersen, and J. Brooks-Gunn, eds., *Encyclopedia of Adolescence*, 637–41. New York: Garland, 1991.

———. "Physical Growth." In P. H. Mussen, ed., *Carmichael's Manual of Child Psychology*, 3rd ed., vol. 1. New York: Wiley, 1970.

Thomas, A., and Chess, S. *The Dynamics of Psychological Development.* New York: Brunner/Mazel, 1980.

———. "The Role of Temperament in the Contributions of Individuals to Their Development." In R. M. Lerner and N. A. Busch-Rossnagel, eds., *Individuals as Producers of Their Own Development: A Life-Span Perspective*, 231–55. New York: Academic Press, 1981.

———. *Temperament and Development.* New York: Brunner/Mazel, 1977.

Thomas, A.; Chess, S.; Birch, H. G.; Hertzig, M. E.; and Korn, S. *Behavioral Individuality in Early Childhood.* New York: New York University Press, 1963.

Thomas, A.; Chess, S.; and Korn, S. J. "The Reality of Difficult Temperament." *Merrill-Palmer Quarterly* 28 (1982): 1–20.

Time-Life Books. *The New Order.* Alexandria, Va.: Time-Life Books, 1989.

Tobach, E. "Evolutionary Aspects of the Activity of the Organism and Its Development." In R. M. Lerner and N. A. Busch-Rossnagel, eds., *Individuals as Producers of Their Own Development: A Life-Span Perspective*, 37–68. New York: Academic Press, 1981.

———. "The Meaning of Cryptanthroparion." In L. Ehrman, G. Omenn, and E. Caspari, eds., *Genetics, Environment, and Behavior*, 219–39. New York: Academic Press, 1972.

———. "Some Evolutionary Aspects of Human Gender." *Journal of Orthopsychiatry* 41 (1971): 710–15.

Tobach, E.; Gianutsos, J.; Topoff, H. R.; and Gross, C. G. *The Four Horsemen: Racism, Sexism, Militarism, and Social Darwinism.* New York: Behavioral Publications, 1974.

Tobach, E., and Greenberg, G. "The Significance of T. C. Schneirla's Contribution to the Concept of Levels of Integration." In G. Greenberg and E.Tobach, eds., *Behavioral Evolution and Integrative Levels,* 1–7. Hillsdale, N.J.: Erlbaum, 1984.

Tobach, E., and Schneirla, T. C. "The Biopsychology of Social Behavior of Animals." In R. E. Cooke and S. Levin, eds., *Biologic Basis of Pediatric Practice,* 68–82. New York: McGraw-Hill, 1968.

Toland, J. *Adolf Hitler.* New York: Doubleday, 1976.

Trivers, R. In *Sociobiology: Doing What Comes Naturally.* A film distributed by Document Associates Inc., New York, N.Y., 1974.

Uphouse, L. L., and Bonner, J. "Preliminary Evidence for the Effects of Environmental Complexity on Hybridization of Rat Brain RNA to Rat Brain DNA." *Developmental Psychobiology* 8 (1975): 171–78.

Volkov, S. "The Written Matter and the Spoken Word." In F. Furet, ed., *Unanswered Questions: Nazi Germany and the Genocide of the Jews,* 33–53. New York: Schocken Books, 1989.

Walsten, D. "Insensitivity of the Analysis of Variance to Heredity-Environment Interaction." *Behavioral and Brain Sciences* 13 (1990): 109–20.

Weinreich, M. *Hitler's Professors: The Part of Scholarship in Germany's Crimes Against the Jewish People.* New York: Yiddish Scientific Institute, 1946.

Williams, G. C. *Adaptation and Natural Selection.* Princeton: Princeton University Press, 1966.

Wilson, E. O. "Academic Vigilantism and the Political Significance of Sociobiology." *BioScience* 183 (1976): 187–90. (Reprinted in A. L. Caplan, ed., *The Sociobiology Debate* [New York: Harper & Row], 291–303.)

———. "A Consideration of the Genetic Foundation of Human Social Behavior." In G. W. Barlow and J. Silverberg, eds., *Sociobiology: Beyond Nature/Nurture,* 295–305. Boulder, Colo.: Westview Press, 1980.

———. "For Sociobiology." *New York Review of Books,* December 11, 1975.

———. "Human Decency Is Animal." *New York Times Magazine,* October 12, 1975.

———. *Sociobiology: The New Synthesis.* Cambridge, Mass.: Harvard University Press, 1975.

Wolfe, R. "Putative Threat to National Security as a Nuremberg Defense for Genocide." *Annals of the American Academy of Political and Social Science* 450 (1980): 46ff.

Ziegler, H. *Die Naturwissenschaft und die Socialdemokratische Theorie.* Stuttgart: Enke, 1893.

INDEX